Essential Mathematics
and
Statistics for Science
Second Edition

Essential Mathematics and Statistics for Science
Second Edition

Graham Currell

Antony Dowman

The University of the West of England, UK

WILEY-BLACKWELL

A John Wiley & Sons, Ltd., Publication

Library of Congress Cataloging-in-Publication Data

Currell, Graham.
  Essential mathematics and statistics for science / Graham Currell, Antony Dowman. – 2nd ed.
    p. cm.
  Includes index.
  ISBN 978-0-470-69449-7 – ISBN 978-0-470-69448-0
  1. Science–Statistical methods. 2. Science–Mathematics. I. Dowman, Antony. II. Title.
  Q180.55.S7C87 2009
  507.2–dc22   2008052795

ISBN: 978-0-470-69449-7 (HB)
     978-0-470-69448-0 (PB)

A catalogue record for this book is available from the British Library.

Typeset in 10/12pt Times and Century Gothic by Laserwords Private Limited, Chennai, India.
Printed and bound in Great Britain by CPI Antony Rowe, Chippenham, Wiltshire.

First Impression   2009

To
Jenny and Felix
Jan, Ben and Jo.

Contents

Preface

The main changes in the second edition have been driven by the authors' direct experience of using the book as a core text for teaching mathematics and statistics to students on a range of undergraduate science courses.

Major developments include:

- Integration of 'how to do it' *video clips* via the Website to provide students with audio-visual worked answers to over 200 'Q' questions in the book.
- Improvement in the *educational development* for certain topics, providing a greater clarity in the learning process for students, e.g. in the approach to handling equations in Chapter 3 and the development of exponential growth in Chapter 5.
- Reorientation in the approach to hypothesis testing to give priority to an understanding of the *interpretation of p-values*, although still retaining the calculation of test statistics. The statistics content has been substantially reorganized.
- Movement of some *content to the Website*, e.g. Bayesian statistics and some of the statistical theory underpinning regression and analysis of variance.
- Revised *computing tutorials* on the Website to demonstrate the use of Excel and Minitab for many of the data analysis techniques. These include *video* demonstrations of the required keystrokes for important techniques.

The book was designed principally as a study text for students on a range of undergraduate science programmes: biological, environmental, chemical, forensic and sports sciences. It covers the majority of mathematical and statistical topics introduced in the first two years of such programmes, but also provides important aspects of experimental design and data analysis that students require when carrying out extended project work in the later years of their degree programmes.

The comprehensive Website actively supports the content of the book, now including extensive video support. The book can be used independently of the Website, but the close integration between them provides a greater range and depth of study possibilities. The Website can be accessed at:

www.wiley.com/go/currellmaths2

The introductory level of the book assumes that readers will have studied mathematics with moderate success to Year 11 of normal schooling. Currently in the UK, this is equivalent to a Grade C in Mathematics in the General Certificate of Secondary Education (GCSE).

There are Revision Mathematics notes available on the associated Website for those readers who need to refresh their memories on relevant topics of basic mathematics – BODMAS, number line, fractions, percentages, areas and volumes, etc. A self-assessment test on these 'basic' topics is also available on the Website to allow readers to assess their need to use this material.

The first eight chapters in the book introduce the *basic* mathematics and statistics that are required for the modelling of many different scientific systems. The remaining chapters are then primarily related to *experimental investigation* in science, and introduce the statistical techniques that underpin data analysis and hypothesis testing.

Over 200 worked Examples in the text are used to develop the various topics. The calculations for many of these Examples are also performed using Microsoft Excel (office.microsoft.com) and the statistical analysis program Minitab (www.minitab.com). The files for these calculations are available via the Website.

Readers can test their understanding as each topic develops by working through over 200 'Q' questions in the book. The numeric answers are given at the end of the book, but full *worked* answers are also available through the Website in both video and printed (pdf) format.

Throughout the book, readers have the opportunity of learning how to use software to perform many of the calculations. This strong integration of paper-based and computer-based calculations both supports an understanding of the mathematics and statistics involved and develops experience with the use of appropriate software for data handling and analysis.

Scientific context

The diverse uses of mathematics and statistics in the various disciplines of science place different emphases on the various topics. However, there is a core of mathematical and statistical techniques that is essentially common to all branches of experimental science, and it is this material that forms the basis of this book. We believe that we have developed a coherent approach and consistent nomenclature, which will make the material appropriate across the various disciplines.

When developing questions and examples at an introductory level, it is important to achieve a balance between treating each topic as pure mathematics or embedding it deeply in a scientific 'context'. Too little 'context' can reduce the scientific interest, but too much can confuse the understanding of the mathematics. The optimum balance varies with topic and level.

The 'Q' questions and Examples in the book concentrate on clarity in developing the topics step by step through each chapter. Where possible we have included a scientific context that is understandable to readers from a range of different disciplines.

Experimental design

The process of good experimental planning and design is a topic that is often much neglected in an undergraduate course. Although the topic pervades all aspects of science, it does not have a clear focus in any one particular branch of the science, and is rarely treated coherently in its own right.

Good experimental design is dependent on the availability of suitable mathematical and statistical techniques to analyse the resulting data. A wide range of such methods are introduced in this book:

- Regression analysis (Chapters 4 and 13) for relationships that are inherently linear or can be linearized.

- Logarithmic and/or exponential functions (Chapter 5) for systems involving natural growth and decay, or for systems with a logarithmic response.
- Modelling with Excel (Chapter 6) for rates of change.
- Probabilities (Chapter 7), frequency and proportions (Chapter 14) and Bayesian statistics (Website) to interpret categorical data, ratios and likelihood.
- Statistical distributions (Chapter 8) for modelling random behaviour in complex systems.
- Statistical analysis (Chapters 9 to 14) for hypothesis testing in a variety of systems.
- Analysis of variance (Chapter 11) for hypothesis testing of complex experimental systems.
- Experimental design overview (Chapter 15).

Computing software

There are various software packages available that can help scientists in implementing mathematics and statistics. Some university departments have strong preferences for one or the other.

Microsoft Excel spreadsheets can be used effectively for a variety of purposes:

- basic data handling – sorting and manipulating data;
- data presentation using graphs, charts, tables;
- preparing data and graphs for export to other packages;
- performing a range of mathematical calculations; and
- performing a range of statistical calculations.

Minitab (Minitab Inc.) is designed specifically for statistical data analysis. The data is entered in columns and a wide range of analyses can be performed using menu-driven instructions and interactive dialogue boxes. The results are provided as printed text, graphs or new column data.

Most students find that the statistical functions in Excel are a helpful *introduction* to using statistics, but for particular problems it is more useful to turn to the packages designed specifically for statistical analysis. Nevertheless, it is usually convenient to use Excel for organizing data into an appropriate layout before exporting to the specialized package.

The book has used Excel 2003 and Minitab 15 to provide all of the software calculations used, and the relevant files are available on the Website. However, there are several other software packages that can perform similar tasks, and information on some of these is also given on the Website.

Most of the graphs in the book have been prepared using Excel, except for those identified as having been produced using Minitab.

On-line Learning Support

The book's Website (www.wiley.com/go/currellmaths2) provides extensive learning support integrated closely with the content of the book.

Important learning elements referenced *within* the book are:

- **Examples** (e.g. **Example 7.12**) with worked answers given directly within the text, and with supporting files available on the Website where appropriate.
- **'Q' questions** (e.g. **Q7.13**) with *numerical* answers at the end of the book, but with *full worked* answers on video or pdf files via the Website.
- **Equations** – referred to using **square brackets**, e.g. **[7.16]**.

The Website for the second edition provides the following structural support:

- **'How to do it' – answers to all 'Q' questions**. Over 200 flash video clips provide worked answers to all of the 'Q' questions in the book, and can be viewed directly over the Internet. The worked answers are also presented in pdf files.
- **Further practice questions**. Additional questions and answers are provided which enable students to further practise/test their understanding. Many students find these particularly useful in some *skill* areas, such as chemical calculations, rearranging equations, logs and exponentials, etc.
- **Excel and Minitab tutorials**. Keystroke tutorials provide a guide to using Excel 2007 and Minitab 15 for some of the important analyses developed in the book.
- **Excel and Minitab files**. These files provide the software calculations for the examples, 'Q' questions, tables and figures presented in the book. In appropriate cases, these are linked with video explanations.
- **Additional materials**. Additional learning materials (pdf files), including revision mathematics (basic skills of the number line, BODMAS, fractions, powers, areas and volumes), Bayesian statistics, transformation of data, weighted and nonlinear regression, data variance.
- **Reference materials**. Statistical tables, Greek symbols.
- **Links**. Access to ongoing development of teaching materials associated with the book, including on-line self-assessment.

Videos

The Website hosts a large number of feedback and instructional videos that have been developed since the first edition of the book was published. Most of these are very short (a few minutes) and provide students with the type of feedback they might expect to receive when asking a tutor 'how to do' a particular question or computer technique. The videos are targeted to produce support just at the point when the student is really involved with trying to understand a particular detailed problem, and provide the focused help that is both required and very welcome.

These videos are used by students of all abilities: advanced students use them just as a quick check on their own self-study, but weaker students can pause and rerun the videos to provide a very effective self-managed 'tutorial'.

The video formats include a 'hand-written' format for paper-based answers, and 'keystroke' demonstrations for computer-based problems. These match directly the form and content of the knowledge and skills that the student is trying to acquire. The separate videos can be viewed directly and quickly over the Internet, using flash technology which is already loaded with most Internet browsers.

1

Mathematics and Statistics in Science

Overview

Science students encounter mathematics and statistics in three main areas:

- Understanding and using theory.
- Carrying out experiments and analysing results.
- Presenting data in laboratory reports and essays.

Unfortunately, many students do not fully appreciate the need for understanding mathematics and/or statistics until it suddenly confronts them in a lecture or in the write-up of an experiment. There is indeed a 'chicken and egg' aspect to the problem:

> Some science students have little enthusiasm to study mathematics until it appears in a lecture or tutorial – by which time it is too late! Without the mathematics, they cannot *fully* understand the science that is being presented, and they drift into a habit of accepting a 'second-best' science *without* mathematics. The end result could easily be a drop of at least one grade in their final degree qualification.

All science is based on a *quantitative* understanding of the world around us – an understanding described ultimately by *measurable* values. Mathematics and statistics are merely the processes by which we handle these quantitative values in an effective and logical way.

Mathematics and statistics provide the network of links that tie together the details of our understanding, and create a sound basis for a fundamental appreciation of science as a whole. Without these quantifiable links, the ability of science to predict and move forward into new areas of understanding would be totally undermined.

In recent years, the data handling capability of information technology has made mathematical and statistical calculations far easier to perform, and has transformed the day-to-day work in many areas of science. In particular, a good spreadsheet program, like Excel, enables both scientists and students to carry out extensive calculations quickly, and present results and reports in a clear and accurate manner.

Essential Mathematics and Statistics for Science 2nd Edition   Graham Currell and Antony Dowman
Copyright © 2009 John Wiley & Sons, Ltd

1.1   Data and Information

Real-world information is expressed in the mathematical world through **data**.

In science, some data values are believed to be fixed in nature. We refer to values that are fixed as **constants**, e.g. the constant c is often used to represent the speed of light in a vacuum, $c = 3.00 \times 10^8 \text{ m s}^{-1}$.

However, most measured values are subject to change. We refer to these values as **variables**, e.g. T for temperature, pH for acidity.

The term **parameter** refers to a variable that can be used to describe a relevant characteristic of a scientific system, or a statistical population (see 7.2.2), e.g. the actual pH of a buffer solution, or the average (mean) age of the whole UK population. The term **statistic** refers to a variable that is used to describe a relevant characteristic of a *sampled* (see 7.2.2) set of data, e.g. five repeated measurements of the concentration of a solution, or the average (mean) age of 1000 members of the UK population.

Within this book we use the convention of printing letters and symbols that represent quantities (constants and variables) in italics, e.g. c, T and p.

The letters that represent units are presented in normal form, e.g. m s^{-1} gives the units of speed in metres per second.

There is an important relationship between data and information, which appears when analysing more complex data sets. It is a basic rule that:

> It is impossible to get more 'bits' of information from a calculation than the number of 'bits' of data that is put into the calculation.

For example, if a chemical mixture contains three separate compounds, then it is necessary to make at least three separate measurements on that mixture before it is possible to calculate the concentration of each separate compound.

In mathematics and statistics, the *number* of bits of information that are available in a data set is called the **degrees of freedom**, df, of that data set. This value appears in many statistical calculations, and it is usually easy to calculate the number of degrees of freedom appropriate to any given situation.

1.2   Experimental Variation and Uncertainty

The uncertainty inherent in scientific information is an important theme that appears throughout the book.

The **true value** of a variable is the value that we would measure if our measurement process were 'perfect'. However, because no process is perfect, the 'true value' is not normally known.

The **observed value** is the value that we produce as our *best estimate* of the true value.

The **error** in the measurement is the difference between the true value and the observed value:

$$\text{Error} = \text{Observed value} - \text{True value} \qquad [1.1]$$

As we do not normally know the 'true value', we cannot therefore know the actual error in any particular measurement. However, it is important that we have some idea of how large the error might be.

The **uncertainty** in the measurement is our *best estimate* of the magnitude of possible errors. The magnitude of the uncertainty must be derived on the basis of a proper understanding of the measurement process involved and the system being measured. The statistical interpretation of uncertainty is derived in 8.2.

The uncertainty in experimental measurements can be divided into two main categories:

Measurement uncertainty. Variations in the actual process of measurement will give some differences when the same measurement is repeated under exactly the same conditions. For example, repeating a measurement of alcohol level in the same blood sample may give results that differ by a few milligrams in each 100 millilitres of blood.

Subject uncertainty. A subject is a representative example of the system (9.1) being measured, but many of the systems in the real world have inherent variability in their responses. For example, in testing the effectiveness of a new drug, every person (subject) will have a slightly different reaction to that drug, and it would be necessary to carry out the test on a wide range of people before being confident about the 'average' response.

Whatever the source of uncertainty, it is important that any experiment must be designed both to counteract the effects of uncertainty and to quantify the magnitude of that uncertainty.

Within each of the two types of uncertainty, *measurement* and *subject*, it is possible to identify two further categories:

Random error. Each subsequent measurement has a random error, leading to *imprecision* in the result. A measurement with a low random error is said to be a *precise* measurement.

Systematic error. Each subsequent measurement has the same recurring error. A systematic error shows that the measurement is *biased*, e.g. when setting the liquid level in a burette, a particular student may always set the meniscus of the liquid a little too low.

The **precision** of a measurement is the best estimate for the purely *random error* in a measurement.

The **trueness** of a measurement is the best estimate for the *bias* in a measurement.

The **accuracy** of a measurement is the best estimate for the *overall error* in the final result, and includes both the effects of a lack of precision (due to random errors) and bias (due to systematic errors).

Example 1.1

Four groups of students each measure the pH (acidity) of a sample of soil, with each group preparing five replicate samples for testing. The results are given in Figure 1.1.

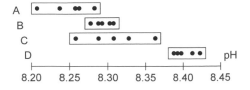

Figure 1.1 Precision and bias in experimental data.

What can be said about the *accuracy* of their results?

It is possible to say that the results from groups A and C show greater *random uncertainty* (less precision) than groups B and D. This could be due to such factors as a lack of care in preparing the five samples for testing, or some electronic instability in the pH meter being used.

Groups B and D show greater precision, but at least one of B or D must have some *bias* in their measurements, i.e. poor 'trueness'. The bias could be due to an error in setting the pH meter with a buffer solution, which would then make every one of the five measurements in the set wrong by the same amount.

With the information given, very little can be said about the overall *accuracy* of the measurements; the 'true' value is not known, and there is no information about possible bias in any of the results. For example *if the true value were pH* = 8.40, this would mean that groups A, B and C were all biased, with the most *accurate* measurement being group D.

The effect of random errors can be managed and quantified using suitable statistical methods (8.2, 8.3 and 15.1.2). The presentation of uncertainty as *error bars* on graphs is developed in an Excel tutorial on the Website.

Systematic errors are more difficult to manage in an experiment, but good experiment design (Chapter 15) aims to counteract their effect as much as possible.

1.3   Mathematical Models in Science

A fundamental building block of both science and mathematics is the *equation*.

Science uses the equation as a *mathematical model* to define the *relationship* between one or more factors in the real world (3.1.6). It may then be possible to use mathematics to investigate how that equation may lead to *new conclusions* about the world.

Perhaps the most famous equation, arising from the general theory of relativity, is:

$$E = mc^2$$

which relates the amount of energy, E (J), that would be released if a mass, m (kg), of matter was converted into energy (e.g. in a nuclear reactor). E and m are both variables and the constant $c(= 3.00 \times 10^8 \text{ m s}^{-1})$ is the speed of light.

Example 1.2

Calculate the amount of matter, m, that must be converted *completely* into energy, if the amount of energy, E, is equivalent to that produced by a medium-sized power station in one year: $E = 1.8 \times 10^{13}$ J.

Rearranging the equation $E = mc^2$ gives:

$$m = \frac{E}{c^2}$$

Substituting values into the equation:

$$m = \frac{1.8 \times 10^{13}}{(3.00 \times 10^8)^2} \Rightarrow 0.000\,20 \text{ kg} \Rightarrow 0.20 \text{ g}$$

This equation tells us that if only 0.20 g of matter is converted into energy, it will produce an energy output equivalent to a power station operating for a year!

This is why the idea of nuclear power continues to be so very attractive.

Example 1.2 indicates some of the common mathematical processes used in handling equations in science: rearranging the equation, using scientific notation, changing of units, and 'solving' the equation to derive the value of an unknown variable.

Equations are used to represent many different types of scientific processes, and often employ a variety of *mathematical functions* to create suitable models.

In particular, many scientific systems behave in a manner that is best described using an exponential or logarithmic function, e.g. drug elimination in the human body, pH values. Example 1.3 shows how both the growth and decay of a bacteria population can be described, in part, by exponential functions.

Example 1.3

Figure 1.2 gives a plot of growth and decay in a bacteria batch colony, by plotting $\log(N)$ against time, t, where N is the number of cells per millilitre.

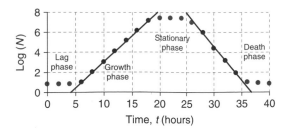

Figure 1.2 Lifecycle of a bacterial population.

The 'straight line' sections of the graph in the 'growth' and 'death' phases are two sections of the lifecycle that can be described by *exponential* functions (5.2).

Another aspect of real systems is that they often have significant inherent *variability*, e.g. similar members of a plant crop grow at different rates, or repeated measurements of the refractive index of glass may give different results. In these situations, we need to develop *statistical models* that we can use to describe the underlying behaviour of the system as a whole.

The particular statistical model that best fits the observed data is often a good guide to the scientific processes that govern the system being measured. Example 1.4 shows the Poisson distribution that could be expected if plants were distributed *randomly* with an average of 3.13 plants per unit area.

Example 1.4

Figure 1.3 shows the numbers (frequencies) of specific plants measured in 100 quadrats of unit area. In sampling the *random* distribution of plants it was found that 22 quadrats had 3 plants, 11 quadrats had 5 plants, etc.

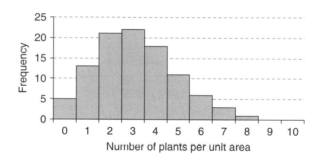

Figure 1.3   Poisson distribution of random plant abundance.

If the distribution of plants were affected by clumping or by competition for survival, then we would expect the *shape of the distribution* to be different.

Excel spreadsheets have become particularly useful for implementing mathematical models of very complex scientific systems. Throughout this book we continue to develop mathematics and statistics in conjunction with their practical applications through Excel.

2
Scientific Data

Overview

- 'How to do it' video answers for all 'Q' questions.
- Revision mathematics notes for basic mathematics: BODMAS, number line, fractions, powers, areas and volumes.
- Excel tutorials: scientific calculations, use of formulae, functions, formatting (scientific numbers, decimal places), etc.

Data in science appears in a variety of forms. However, there is a broad classification of data into *two* main categories:

- **Quantitative data.** The numeric value of quantitative data is recorded as a *measurable (or parametric) variable*, e.g. time, pH, temperature, etc.
- **Qualitative (or categorical) data.** Qualitative data is *grouped* into different classes, and the *names* of the classes serve only to distinguish, or rank, the different classes, and have no other quantitative value, e.g. grouping people according to their nationality, eye colour, etc.

Quantitative data can be further divided into:

Discrete data. Only *specific values* are used, e.g. counting the number of students in a class will only give *integer* values.

Continuous data. Using values specified to any accuracy as required, e.g. defining time, using seconds, to any number of decimal places as appropriate (e.g. 75.85206 s).

Quantitative data can be further subdivided into:

Ratio data. The 'zero' of a ratio scale has a true 'zero' value in science, and the ratios of data values also have scientific meaning. For example, the zero, 0 K, of the *absolute temperature scale* in thermodynamics is a *true* 'absolute zero' (there is nothing colder!), and 100 K is twice the absolute 'temperature' of 50 K.

Interval data. The 'zero' of an interval scale does *not* have a true 'zero' value in science, and the ratios of data values do not have scientific meaning. Nevertheless the data *intervals* are still significant. For example, the zero, $0\,^\circ\text{C}$, of the *Celsius temperature scale*, is just the temperature of melting ice and not a true 'zero', and $100\,^\circ\text{C}$ is not 'twice as hot' as $50\,^\circ\text{C}$. However, the degree intervals are the same in both the absolute and Celsius scales.

Qualitative data can be further divided into:

Ordinal data. The classes have a sense of *progression* from one class to the next, e.g. degree classifications (first, upper second, lower second, third), opinion ratings in a questionnaire (excellent, good, satisfactory, poor, bad).

Nominal (named) data. There is *no* sense of progression between classes, e.g. animal species, nationality.

This chapter is concerned mainly with calculations involving *continuous quantitative* data, although examples of other types of data appear elsewhere within the book. The topics included relate to some of the most common calculations that are performed in science:

- Using scientific (or standard) notation.
- Displaying data to an appropriate precision.
- Handling units, and performing the conversions between them.
- Performing routine calculations involving chemical quantities.
- Working with angular measurements in both degrees and radians.

2.1 Scientific Numbers

2.1.1 Introduction

This unit describes some of the very common arithmetical calculations that any scientist needs to perform when working with numerical data. Students wishing to refresh their memory of basic mathematics can also refer to the revision resources available on the book's dedicated Website.

2.1.2 Scientific (standard) notation

Scientific notation is also called *standard notation* or *exponential notation*.

In **scientific notation**, the digits of the number are written with the most significant figure before the decimal point and all other digits after the decimal point. This 'number' is then multiplied by the correct 'power of 10' to make it equal to the desired value. For example:

$$230 \quad = 2.30 \times 10^2$$

$$0.00230 = 2.30 \times 10^{-3}$$

$$2.30 \quad = 2.30 \times 10^0 \Rightarrow 2.30 \times 1 \Rightarrow 2.30$$

In Excel, other software and some calculators, the 'power of 10' is preceded by the letter 'E', e.g. the number -3.56×10^{-11} would appear as $-3.56\text{E-}11$.

In a calculator, the 'power of 10' is entered by pressing the '$\times 10^x$' or 'EXP' button, e.g. entering 2.6×10^3 by using the keystrokes [2][.][6][$\times 10^x$][3].

Q2.1

Express the following numbers in scientific notation:

(i) 42600   (v) 0.045×10^4

(ii) 0.00362   (vi) 26.6×10^3

(iii) 10000   (vii) 3.2E3

(iv) 0.0001   (viii) 4.5E-6

2.1.3 Multiplying (dividing) in scientific notation

When multiplying (or dividing) in scientific notation, it is possible to multiply (or divide) the numbers *separately* and add (or subtract) the 'powers of 10', as in the next example.

Example 2.1

Multiplication in scientific notation:

$$4.2 \times 10^3 \times 2.0 \times 10^4 \Rightarrow \quad (4.2 \times 2.0) \times (10^3 \times 10^4) \quad \Rightarrow \quad 8.4 \times 10^{3+4} \quad \Rightarrow 8.4 \times 10^7$$

Separating numbers and powers   Adding powers

Division in scientific notation:

$$\frac{4.2 \times 10^3}{2.0 \times 10^4} \Rightarrow \quad \frac{4.2}{2.0} \times \frac{10^3}{10^4} \quad \Rightarrow \quad 2.1 \times 10^{3-4} \quad \Rightarrow 2.1 \times 10^{-1*}$$

Separating numbers and powers   Subtracting powers

*For simplicity of presentation, 2.1×10^{-1} would normally be written just as 0.21.

It is often necessary to 'adjust' the position of the decimal point (and 'power of 10') to return the final number to true scientific notation, as in the final step in Example 2.2.

Example 2.2

A simple multiplication gives:

$$4.0 \times 10^5 \times 3.5 \times 10^{-3} = 4.0 \times 3.5 \times 10^5 \times 10^{-3} \Rightarrow 14 \times 10^{5+(-3)} \Rightarrow 14 \times 10^2$$

However, the result is not in *scientific notation*, and should be adjusted to give:

$$14 \times 10^2 = 1.4 \times 10^3$$

Q2.2

Evaluate the following, giving the answers in scientific notation (calculate 'by hand' and then check the answers on a calculator):

(i) 120000×0.003

(ii) $5.0 \times 10^5 \times 3.0 \times 10^{-3}$

(iii) $\dfrac{1.2 \times 10^5}{3.0 \times 10^3}$

(iv) $4500 \div 0.09$

(v) $0.0056 \times 4.0 \times 10^3$

(vi) $\dfrac{1.2 \times 10^5}{3.0 \times 10^{-3}}$

2.1.4 Adding (subtracting) in scientific notation

Before adding or subtracting scientific numbers it is important to get both numbers to the same 'power of 10'.

It is then possible to simply add (or subtract) the numbers.

Example 2.3

To add 3.46×10^3 to 2.120×10^4 we first change 3.46×10^3 to 0.346×10^4 so that both numbers have the multiplier '$\times 10^4$'.

We can then write:

$$3.46 \times 10^3 + 2.120 \times 10^4 = 0.346 \times 10^4 + 2.120 \times 10^4 \Rightarrow (0.346 + 2.120) \times 10^4$$
$$\Rightarrow 2.466 \times 10^4$$

Similarly, to subtract 2.67×10^{-2} from 3.0×10^{-3} we first change 2.67×10^{-2} to 26.7×10^{-3} so that both numbers have the multiplier '$\times 10^{-3}$', and we can then write:

$$3.0 \times 10^{-3} - 2.67 \times 10^{-2} = 3.0 \times 10^{-3} - 26.7 \times 10^{-3} \Rightarrow (3.0 - 26.7) \times 10^{-3}$$
$$\Rightarrow -23.7 \times 10^{-3} \Rightarrow -2.37 \times 10^{-2}$$

Note that the answer should be left in correct scientific notation form.

Q2.3

Evaluate the following, giving the answers in scientific notation (calculate 'by hand' and then check the answers on a calculator):

(i) $1.2463 \times 10^3 - 42.1$ (ii) $\dfrac{7.2463 \times 10^6 - 1.15 \times 10^5}{3.0 \times 10^{-3}}$

2.1.5   Significant figures (sf)

The **most significant figure** (or digit) in a number is the first non-zero number reading from the left, e.g. '4' in each of the numbers 456 and 0.047.

The **least significant figure** (or digit) is the last digit to the right whose value is considered to carry valid information.

Example 2.4

According to the 1951 Census, the population of Greater London was 8346137. If I state that the population was 8350000, correct to 3 *significant figures (sf)*, then I am claiming (correctly) that the population was closer to 8350000 than to either 8340000 or 8360000.

The figure '8' is the *most* significant figure, and the '5' is the *least* significant figure. The zeros are included to indicate the appropriate 'power of 10'.

After the decimal point, a final zero should only be included if it is *significant*. For example:

3.800 to 4 sf would be written as 3.800
3.800 to 3 sf would be written as 3.80
3.800 to 2 sf would be written as 3.8

The number of significant figures chosen will depend on the precision or accuracy with which the value is known.

2.1.6   Decimal places (dp)

The **format** of numbers can be specified by defining how many **decimal places (dp)** are included after the decimal point. For example, 9.81 $\mathrm{m\,s}^{-2}$ is the acceleration due to gravity written to 2 decimal places.

2.1.7 Rounding numbers

It is important, when information is presented in the form of data, that the data is an accurate representation of the information. There is uncertainty in all scientific 'information' (1.2), and the number of significant figures used in displaying the data should not imply a greater precision than is actually the case. For example, it would not be correct to quote an answer as 1.145917288 simply because the calculator displayed that many digits – it is exceedingly rare for any scientific measurement to be that precise (± 0.000000001)!

To get the right number of significant figures (sf) or decimal places (dp), it is sometimes necessary to 'round off' the number to the nearest value.

When rounding numbers to specific interval values, any number that is *more than halfway* between values will round *up* to the next value, and any number *less than halfway* will round *down*.

Example 2.5

Rounding:

(i)   70860 to 3 sf gives 70900

(ii) 70849 to 3 sf gives 70800

(iii) 5.6268×10^{-3} to 4 sf gives 5.627×10^{-3}

(iv) 3.194 to 2 dp gives 3.19

(v) 3.196 to 2 dp gives 3.20

If the number is *exactly halfway* between values, it is common practice (including rounding in Excel) that the *halfway* value *always* rounds upwards. However, it is sometimes claimed that, for the halfway value, the number *should* round so that the *last digit is even*.

Example 2.6

(i)   Rounding 70550 to 3 sf gives 70600

(ii) Rounding 70850 to 3 sf *normally* gives 70900

(iii) Rounding 0.275 to 2 dp gives 0.28

(iv) Rounding 3.185 to 2 dp *normally* gives 3.19

Q2.4

Round the following numbers to the required numbers of significant figures (sf) as stated:

(i)   0.04651 to 2 sf

(ii) 0.04649 to 2 sf

(iii) 13.97 to 3 sf

(iv) 7.3548×10^3 to 3 sf

(v)   26962 to 3 sf

(vi) 11.250 to 3 sf

(vii) 11.150 to 3 sf

(viii) 5.6450×10^{-3} to 3 sf

Q2.5

Round the following numbers to the required numbers of decimal places (dp) as stated:

(i) 0.04651 to 3 dp   (iii) 426.891 to 2 dp
(ii) 7.9999 to 2 dp   (iv) 1.3450 to 2 dp

When presenting a final calculated value, the number of significant figures or decimal places should reflect the accuracy of the result. Simply performing a mathematical calculation cannot improve the overall accuracy or precision of the original information.

Q2.6

Add the following masses and give the result to an *appropriate* number of decimal places (hint: in this case the total value cannot have more decimal places than the *least precise* of all the separate masses):

$$0.643 \text{ g}, 3.10 \text{ g}, 0.144 \text{ g}, 0.0021 \text{ g}$$

It is also important that the rounding process should not be applied until the *end of the calculation*. If the data is rounded too early, then it is quite possible that the small inaccuracies created will be magnified by subsequent calculations. This may result in a final error that is much greater than any uncertainty in the real information.

2.1.8   Order of magnitude

If a value increases by one 'order of magnitude', then it increases by (very) *approximately 10 times*:

- one order of magnitude is an increase of *10* times;
- two orders of magnitude is an increase of 100 times; and
- an increase of 100000 times is five orders of magnitude.

Example 2.7

What are the differences in 'orders of magnitude' between the following pairs of numbers?

(i) 46800 and 45   (ii) 5.6 mm and 3.4 km

Answers:

  (i) 46800 is three *orders of magnitude* greater than 45

  (ii) 5.6 mm is six *orders of magnitude* less than 3.4 km

2.1.9   Estimations

It is often useful to check complicated calculations by carrying out simple calculations 'by hand' using values approximated to 1 (or 2) significant figures.

Example 2.8

If my calculation suggests that 0.4378×256.2 gives the answer 1121.6436, I can check the result as follows.

Replace the numbers by approximate values 0.4 and 300, and multiply them 'by hand' to get $0.4 \times 300 = 120$.

I then find out that my calculated answer is one order of magnitude out – I have put the decimal point in the wrong place, and the correct answer should be 112.16436.

Q2.7

Estimate, without using a calculator, the approximate speed (in miles per hour) of an aeroplane that takes 4 hours and 50 minutes to fly a distance of 2527 miles.

Is the answer likely to be too high or too low?

2.1.10   Using a calculator

The most appropriate hand calculator for the science student should be inexpensive, easy to use, and have a basic scientific capability. This capability should include logarithms, the exponential function (e), trigonometric functions, the use of brackets, and basic statistical calculations (mean, standard deviation, etc). The more expensive and sophisticated calculators (e.g. with graphics) should be avoided unless the student is confident in how to use them.

Q2.8

Use a calculator to evaluate the following expressions:

  (i) $1/(2.5 \times 10^4)$ Use the reciprocal key, '1/x' or 'x^{-1}'

(ii) $(-0.0025) \div (-1.2 \times 10^{-6})$ Use the '$\times 10^x$' or 'EXP' key for the power of 10

(iii) $3.2^{-1.6}$ Use the key 'x^y' or '$\wedge$'

(iv) 3.487^2 Use the key 'x^2'

(v) Square root of 0.067   Use the square root key '$\sqrt{}$'

2.2 Scientific Quantities

2.2.1 Introduction

Quantitative measurements are made in relation to agreed 'units' of quantity. For example, the distances for Olympic races are expressed as multiples of an agreed 'unit' of distance (the *metre*): 100 *metres*, 400 *metres*, 1500 *metres*, etc.

The handling of 'units' should be a simple process. However, some students try to work out the conversion of units 'in their heads', and get confused with multiple multiplications and divisions. The answer is to break up the problem into a number of very simple steps, writing down each step in turn.

2.2.2 Presenting mixed units

Most people are very familiar with common 'mixed' units such as miles per hour for speed or pounds per month for wages. However, when writing out such units in full, using the word 'per' takes up a lot of space, and in science it is more convenient to use abbreviated forms. For example, speed is calculated by *dividing* distance by time, and consequently the units become metres *divided* by seconds: m/s or m s^{-1}. However, the format using the oblique '/' for 'per' (e.g. 'm/s') *should not be used* for units, and should be replaced by formats with negative powers, e.g. 'm s^{-1}'.

The units of a mixed variable represent the *process* used to calculate the value of that variable. Some examples of equivalent forms are given below:

Variable	Units	Unit format
Speed	metres per second	m s^{-1}
Density	kilograms per cubic metre	kg m^{-3}
Pressure	newtons per square metre	N m^{-2}

By convention, units are shown in normal (not italic) font with a space between each subunit. Where a unit is derived from a person's name, the first letter of the unit's name is given in lower case, although the unit is give a capital letter, e.g. 1 newton is written as 1 N.

2.2.3 SI units

SI (Système International) units derive from an international agreement to use a common framework of units, which is based on a set of seven fundamental units, as in Table 2.1:

Table 2.1. Fundamental SI units.

SI unit	Symbol	Measures	Defined using:
kilogram	kg	Mass	standard platinum–iridium mass
second	s	Time	oscillations of a caesium-137 atom
metre	m	Length	distance travelled by light in a fixed time
kelvin	K	Temperature	temperature of triple point of water
mole	mol	Amount	comparison with 0.012 kg of carbon-12
ampere	A	Electric current	force generated between currents
candela	cd	Light output	intensity of a light source

Other units are derived as combinations of the fundamental units. Some examples are given in Table 2.2.

Table 2.2. Derived SI units.

SI unit	Symbol	Measures	Equivalence to fundamental units
newton	N	Force	$1\,\text{N} = 1\,\text{kg m s}^{-2}$
joule	J	Energy	$1\,\text{J} = 1\,\text{N m}$
watt	W	Power	$1\,\text{W} = 1\,\text{J s}^{-1}$
pascal	Pa	Pressure	$1\,\text{Pa} = 1\,\text{N m}^{-2}$
hertz	Hz	Frequency	$1\,\text{Hz} = 1 \text{ cycle per second} = 1\,\text{s}^{-1}$

Various prefixes are used to magnify or reduce the size of a particular unit according to the 'power of 10' ratios in Table 2.3.

Table 2.3. Powers of 10 in SI units.

Power of 10	10^9	10^6	10^3	10^{-3}	10^{-6}	10^{-9}	10^{-12}
Name	giga-	mega-	kilo-	milli-	micro-	nano-	pico-
Prefix	G	M	k	m	μ	n	p

Note that 'centi-' is a common prefix for the power 10^{-2} (i.e. *one-hundredth*) and 'deci-' for the power 10^{-1} (i.e. *one-tenth*), but neither are true SI units.

Q2.9

 By how many 'orders of magnitude' (see 2.1.8) is 4.7 km (kilometres) larger than 6.2 nm (nanometres)?

2.2.4   Conversion of units

The examples below, Example 2.9 to Example 2.15, show that:

- Conversion should be performed in *easy* stages, making *simple* changes at each stage.
- It often helps to write out any complex units as statements in *words*.

For example:	**If x units of A**	**are equivalent to**	**y units of B**
Dividing both sides by x:	**1 unit of A**	**is equivalent to**	**$\dfrac{y}{x}$ units of B**
Multiplying both sides by z:	**z units of A**	**are equivalent to**	**$z \times \dfrac{y}{x}$ units of B**

Example 2.9

If 5.000 miles are equal to 8.045 km, convert 16.3 miles into kilometres.

Starting with:	5.000 miles are equal to	8.045 km
Divide both sides by 5:	1.000 mile is equal to	$\dfrac{8.045}{5.000} = 1.609$ km
Multiply both sides by 16.3:	16.3 miles are equal to	$16.3 \times \dfrac{8.045}{5.000} = 26.2$ km

---

It is important to be able to calculate **reciprocal** conversions, as follows.

Example 2.10

Examples of taking the reciprocals of unit conversions:

1.0 m is equivalent to 100 cm $= 1.00 \times 10^2$ cm

Hence: 1.0 cm is equivalent to $\dfrac{1}{100}$ m $= 0.01$m $\Rightarrow 1.00 \times 10^{-2}$ m

1.000 mile is equivalent to 1.609 km

Hence: 1.000 km is equivalent to $\dfrac{1}{1.609}$ miles $= 0.622$ miles

1.000 L is equivalent to 1000 mL $= 1.000 \times 10^3$ mL

Hence: 1.0 mL is equivalent to $\dfrac{1}{1000}$ L $= 0.001$ L $\Rightarrow 1.0 \times 10^{-3}$ L

and: 10 mL is equivalent to $\dfrac{10}{1000}$ L $= 0.01$L $\Rightarrow 1.0 \times 10^{-2}$ L

Units are often used in a 'power' form, e.g. the units of area are m^2.
Conversion factors will also be raised to the **same power as the unit**.

Example 2.11

Examples using *powers* of units:

Distance: $1.0 \text{ m} = 100 \text{ cm} \Rightarrow 1.0 \times 10^2 \text{ cm}$

Area: $1.0 \text{ m}^2 = 100 \times 100 \text{ cm}^2 \Rightarrow 1.0 \times 10^2 \times 1.0 \times 10^2 \text{ cm}^2 \Rightarrow 1.0 \times 10^4 \text{ cm}^2$

Volumes: $1.0 \text{ m}^3 = 100 \times 100 \times 100 \text{ cm}^3 \Rightarrow 1.0 \times 10^2 \times 1.0 \times 10^2 \times 1.0 \times 10^2 \text{ cm}^3$

$\Rightarrow 1.0 \times 10^6 \text{ cm}^3$

$$1.0 \text{ cm}^3 = \frac{1}{1.0 \times 10^6} \text{ m}^3 \Rightarrow 1.0 \times 10^{-6} \text{ m}^3$$

Distance: $1.000 \text{ mile} = 1.609 \text{ km}$

Areas: $1.000 \text{ square mile} = 1.609 \times 1.609 \text{ km}^2 \Rightarrow 1.609^2 \text{ km}^2 \Rightarrow 2.589 \text{ km}^2$

$$1 \text{ km}^2 = \frac{1}{2.589} \text{ square miles} \Rightarrow 0.386 \text{ square miles}$$

Example 2.12

Express a volume of 30 mm^3 in units of m^3.

The first step is to start from what is known, i.e. 1.0 m = 1000 mm:

- then a cubic metre is the volume of a cube with each side of length 1000 mm
- *hence the volume of a cubic metre*, $1 \text{ m}^3 = 1000 \times 1000 \times 1000 \text{ mm}^3 \Rightarrow 1.0 \times 10^9 \text{ mm}^3$
- then $1 \text{ mm}^3 = \dfrac{1}{1.0 \times 10^9} \text{ m}^3 \Rightarrow 1.0 \times 10^{-9} \text{ m}^3$
- and $30 \text{ mm}^3 = 30 \times 1.0 \times 10^{-9} \text{ m}^3 \Rightarrow 30 \times 10^{-9} \text{ m}^3 \Rightarrow 3.0 \times 10^{-8} \text{ m}^3$.

With mixed units, **convert each unit separately** in a step-by-step conversion.

Example 2.13

Express a speed of 9.2 mph in units of $m\,s^{-1}$, given that 1.0 km = 0.6215 miles.

Start from what is known, i.e. 1.0 km is equal to 0.6215 miles:

- The reciprocal conversion: $1.0 \text{ mile} = \dfrac{1}{0.6215} \text{ km} \Rightarrow 1.609 \text{ km} \Rightarrow 1609 \text{ m}$.

- 9.20 miles $= 9.20 \times 1609$ m $\Rightarrow 14802$ m.
- 1 hour $= 3600$ seconds.
- *A speed of 9.2 miles per hour means 9.2 miles are travelled in 1 hour.*
- This is the same as 14802 m travelled in 3600 seconds.
- Speed $=$ Distance travelled in each second $\Rightarrow \dfrac{14802}{3600}$ m s$^{-1} \Rightarrow 4.11$ m s^{-1}.

In a complex conversion, it is useful to convert equivalence equations to **unit values** before calculating a new value. This is illustrated in Example 2.14.

Example 2.14

A lysozyme solution has 15 enzyme units of activity in 22 mL of solution. Calculate the number of enzyme units in 100 mL.

Start from what is known:

22 mL of solution contains 15 enzyme units

Convert to *unit value* of 1 mL:

1 mL of solution contains $\dfrac{15}{22}$ enzyme units

Taking the new value of 100 mL:

100 mL of solution contains $100 \times \dfrac{15}{22} = 68.182$ enzyme units

Example 2.15

A lysozyme solution has 15 enzyme units of activity in 22 mL of solution. If the concentration of protein in the solution is 0.5 grams per 100 mL, calculate the specific lysozyme activity of the solution in enzyme units per milligram of protein.

In this case it would be useful to convert to a *common volume* of 100 mL:

- From Example 2.14 we know that 100 mL of solution contains 68.182 enzyme units.
- We know also that 100 mL of solution contains 0.5 g protein.
- Hence 0.5 g protein is equivalent to an activity of 68.182 units.
- 1.0 g protein will be equivalent to an activity of $\dfrac{68.182}{0.5} = 136.4$ units.
- Activity (per gram of protein) $= 136.4$ units per gram.
- Activity (per milligram of protein) $= \dfrac{136.4}{1000} \Rightarrow 0.1364$ units per milligram.

Q2.10

	Convert:	Use conversion:
(i)	nutrition energy of 750 kcal (Cal) to kilojoules (kJ),	1 kcal = 4.2 kJ
(ii)	nutrition energy of 1200 kilojoules (kJ) to kcal (Cal),	1 kcal = 4.2 kJ
(iii)	a mass of 145 pounds (lb) to kilograms (kg),	1 kg = 2.20 lb
(iv)	cross-section of a plank of wood 6 inches by 1 inch into millimetres (mm),	1 inch = 25.4 mm
(v)	a volume of 5 UK gallons into litres (L),	1 gallon = 8 pints 1 pint = 568 mL
(vi)	What weight (mass) of protein has a nutrition energy value of 46 kcal?	1 g of protein = 4 kcal

Q2.11

(i) How many hectares are there in 1.0 km^2? (1 hectare = 1×10^4 m^2)

(ii) If the density of iron is 7.9 $g\,cm^{-3}$ calculate the density in units of kg m^{-3}.

(iii) A fertilizer is to be spread at the rate of 0.015 $g\,cm^{-2}$. What is the spreading rate in $kg\,m^{-2}$?

(iv) What is a petrol consumption of 40 miles per gallon (mpg) in litres per 100 kilometres? (1 mile = 1.61 km and 1 gallon = 4.55 litres)

2.3 Chemical Quantities

2.3.1 Introduction

Calculations involving chemical quantities are needed across a range of different scientific disciplines.

This unit aims to clarify the basic relationships between the two main ways of measuring **quantity** in chemical calculations:

- **mass** of material; or
- **numbers** of molecules/atoms.

This then leads to consideration of the concentration of chemical solutions.

2.3.2 Quantity (grams and moles)

The standard unit of **mass** is the kilogram (kg), and we also use grams (g), milligrams (mg), micrograms (μg). For example, we may buy 1 kg of salt from a shop, or weigh out 10 g of sodium chloride in a laboratory.

However, it is also common to measure quantity by **number**. We often need a measure of **number** to buy integer (whole) numbers of items such as eggs, oranges or buns.

When shopping, a common unit of number is the 'dozen', where:

- One **dozen** of any item = 12 items.

For example, buying half a dozen eggs = 0.5 dozen $\Rightarrow$ 0.5 × 12 $\Rightarrow$ 6 eggs

We also need a measure of **number** in chemistry because when atoms and molecules react, they do so in simple whole (integer) numbers.

We know that *one* water molecule, H_2O, contains *two* hydrogen atoms, H, plus *one* oxygen atom, O:

$$H_2O \Leftrightarrow 2H + O$$

However, when dealing with atoms and molecules in chemistry, a 'dozen' is far too small a quantity, and instead we **count** atoms and molecules using the much larger 'mole':

- 1 **mole** of any item $\Rightarrow$ 6.02 × 10^{23} items (to 3 significant figures).

For example, weighing out 0.5 moles of sodium chloride (NaCl) gives 0.5 × 6.02 × 10^{23} $\Rightarrow$ 3.01 × 10^{23} molecules of sodium chloride:

---

1 mole of *any* substance will contain the *same number* (= 6.02 × 10^{23}) of items

[2.1]

The **Avogadro constant** is the number of items in 1 *mole* of *any* substance:

$$N_A = 6.02 \times 10^{23} \text{ mol}^{-1} \quad \text{(to 3 sf)}$$

Counting molecules and atoms, we can describe the formation of water:

1 mole of H_2O molecules $\Leftrightarrow$ **2 moles** of H atoms + **1 mole** of O atoms

The above statement using 'moles' gives a clearer understanding of the **chemical formation, H_2O,** of water than the equivalent statement using 'mass':

18 g of water $\Leftrightarrow$ **2 g** of hydrogen + **16 g** of oxygen

---

Example 2.16

Calculate the mass of 1 mole of hydrogen **molecules**, H_2, given that the mass of 1 mole of hydrogen **atoms**, H, is 1.0 g (to 2 sf).

1 mole of H_2 *molecules* consists of **2 moles** of H *atoms*.

Hence, the mass of 1 mole of hydrogen molecules, H_2, is 2×1.0 g $= 2.0$ g.

2.3.3 Relative atomic and molecular masses, A_r and M_r

A key calculation in chemistry involves working out the mass required of a substance to obtain a given number of moles of that substance. As illustrated in Example 2.16, this conversion depends on the ratio of the mass of a single molecule of the compound to a mass (approximately) equal to that of a single hydrogen atom:

- **Relative atomic mass**, A_r (also written *RAM*), is used for the relative mass of an element, and is equal to the *ratio* of the *average* mass of 1 atom of that element to a mass equal (almost) to 1 hydrogen atom. On this basis:

$$A_r \text{ for hydrogen, H } = 1.0 \text{ (to 1dp)}$$
$$A_r \text{ for oxygen, O } = 16.0 \text{ (to 1 dp)}$$
$$A_r \text{ for carbon, C } = 12.0 \text{ (to 1 dp)}$$

- **Relative molecular mass**, M_r (also called **molecular weight** or written *RMM*), of a substance is equal to the *ratio* of the *average* mass of 1 molecule of that substance to a mass equal (almost) to 1 hydrogen atom. On this basis:

M_r for hydrogen, H_2	$= 2 \times 1.0$	$= 2.0$ (to 1 dp)
M_r for water, H_2O	$= 2 \times 1.0 + 16.0$	$= 18.0$ (to 1 dp)
M_r for methane, CH_4	$= 12.0 + 4 \times 1.0$	$= 16.0$ (to 1 dp)

The **exact** values for A_r and M_r are actually based on the ratio of the atomic and molecular masses to one-twelfth of the mass of the carbon-12 isotope (written ^{12}C). Using this scale the mass of 1 mole of H atoms equals 1.01 g (and not 1.00 g). However, in all but the most exact calculations, it is still useful to think of the scale of masses starting with H = 1.0, at least to 1 decimal place.

The term **average** mass is used to allow for the mixture of isotopes of different masses that occurs for all elements, as given in Example 2.17.

Example 2.17

In a naturally occurring sample of chlorine atoms, 76 % of them will be the ^{35}Cl isotope (with $A_r = 35.0$) and approximately 24 % will be the ^{37}Cl isotope (with $A_r = 37.0$).

Calculate the *average* A_r for the mixture.

Taking 100 atoms of naturally occurring chlorine, 76 will have a 'mass' = 35.0 and the remainder a 'mass' = 37.0.

Total 'mass' for 100 atoms = $76 \times 35.0 + 24 \times 37.0 \Rightarrow 3548$.

Average 'mass' in a natural sample of chlorine, $A_r = 3548/100 \Rightarrow 35.5$ (to 1 dp)

Q2.12

In a naturally occurring sample of boron atoms about 80 % will be the ^{11}B isotope (with $A_r = 11.0$) and approximately 20 % the ^{10}B isotope (with $A_r = 10.0$). Estimate the *average* A_r for the mixture.

It is also useful to define the mass (in grams) of 1 mole of the substance:

- **Molar mass, M_m,** of a substance is the mass, *in grams*, of 1 *mole* of that substance. Units are g mol^{-1}.

This now gives us the key statement that links the measurement by mass (in grams) with the number of moles of any substance:

---

1 mole of a substance has a mass in grams (molar mass) numerically

equal to the value of its relative molecular mass, M_r   [2.2]

In practice, the relative molecular mass, M_r, for a molecule is calculated by adding the relative atomic masses, A_r, of its various atoms.

Example 2.18

Calculate the relative molecular mass, molar mass and mass of 1 mole of calcium carbonate, $CaCO_3$, given relative atomic masses $Ca = 40.1$, $C = 12.0$, $O = 16.0$ (all values to 1 dp).

$$M_r = 40.1 + 12.0 + 3 \times 16.0 \Rightarrow 100.1 \,(\text{a pure number})$$

Molar mass $= 100.1 \text{ g mol}^{-1}$ (equals the relative molecular mass, M_r, in grams per mole).

Mass of 1 mole $= 100.1$ g (numerically equals the molar mass in grams).

Q2.13

Using relative atomic masses $C = 12.0$, $H = 1.0$, $O = 16.0$, calculate the following values for aspirin, $C_9H_8O_4$, giving the relevant units:

(i) relative molecular mass, M_r
(ii) molar mass, M_m
(iii) mass of 1 mole

2.3.4 Conversion between moles and grams

Consider a substance, X, with a relative molecular mass, M_r (for an element, we use relative atomic mass, A_r, instead of M_r).

From equation [2.2]:

• 1 mole of X has a mass of M_r g.

Hence, for n moles of the substance:

• n moles of X has a mass of $n \times M_r$ g.

If n moles of the substance has a mass m g, we can write:

$$m = n \times M_r \qquad (m \text{ in grams}) \qquad [2.3]$$

We can rearrange the equation by dividing m by M_r on the left-hand side (LHS), leaving n on the right-hand side (RHS), and then swapping sides to give:

$$n = \frac{m}{M_r} \quad (m \text{ in grams}) \qquad [2.4]$$

These two equations allow us to perform simple conversions between the quantity of a substance measured in grams, m, and the same quantity measured in numbers of moles, n.

Example 2.19

Calculate the following for sodium hydroxide, NaOH ($M_r = 40$):

(i) mass (in g) of 1 mol of NaOH
(ii) mass (in g) of 0.4 mol of NaOH
(iii) number of moles of NaOH that has a mass of 1.0 g
(iv) number of moles of NaOH that has a mass of 8.0 g

Answers:

(i) 1 mol of NaOH has a mass of 40 g (from the definition of a 'mole')
(ii) 0.4 mol of NaOH has a mass $m = n \times M_r \Rightarrow 0.4 \times 40 \Rightarrow 16$ g
(iii) no. of moles $n = \dfrac{m}{M_r} \Rightarrow \dfrac{1}{40} \Rightarrow 0.025$ mol

(iv) no. of moles $n = \dfrac{m}{M_r} \Rightarrow \dfrac{8}{40} \Rightarrow 0.20$ mol

---

Q2.14

Calculate the following for sodium carbonate, Na_2CO_3 ($M_r = 106$):

(i) mass (in g) of 1 mol of Na_2CO_3
(ii) mass (in g) of 0.15 mol of Na_2CO_3
(iii) number of moles of Na_2CO_3 that has a mass of 3.5 g

Q2.15

A sample of benzoic acid with a mass of 2.2 g was found, by titration, to be an amount equal to 0.018 moles. Calculate:

(i) molar mass
(ii) relative molecular mass

2.3.5 Concentration

The concentration of a solution is the amount of solute per unit volume of solution.
    The basic unit of volume, m^3, is a large unit, and it is common to use the smaller:

- litre, L (which equals a cubic decimetre, dm^3); or
- cm^3 (sometimes written as cubic centimetres, cc).

The symbol for the litre should normally be written as lower case l. However, in print, this
can be easily confused with the number 1, or with the upper case I, and we have opted to use
upper case L to avoid such confusion:

$$1\ L = 1\ dm^3 \Rightarrow 1000\ cm^3 \Rightarrow 1 \times 10^{-3}\ m^3 \qquad [2.5]$$

We now consider that n moles of solute X has a mass, m g, and is dissolved in a solution that
occupies a volume, V litres. The concentration of the solution is defined as:

$$\text{Concentration} = \frac{\text{Quantity of solute}}{\text{Volume of solution}}$$

There are *two* primary ways of recording the concentration, C, of a solution, and the form of
the equation depends on the units used to express the concentration:

- **Concentration in grams per litre (g L^{-1})**

$$C(\text{in g L}^{-1}) = \frac{m}{V} \qquad [2.6]$$

- **Molar concentration (mol L^{-1})** (also called **molarity, M**) is the number of moles per litre

$$C(\text{in mol L}^{-1}) = \frac{n}{V} \qquad [2.7]$$

Note that $1.0\ mol\ L^{-1}$ can also be written as $1.0\ mol\ dm^{-3}$ or 1.0 M.

Example 2.20

It is often necessary to change the units in a volume calculation, e.g.

(i)   1.0 mL $\Rightarrow$ 1.0/1000 L $\Rightarrow$ 0.001 L $\Rightarrow$ 1.0×10^{-3} L

(ii) 250 mL $\Rightarrow$ 250/1000 L $\Rightarrow$ 0.25 L

(iii) 0.75 mL $\Rightarrow$ 0.75/1000 L $\Rightarrow$ 7.5×10^{-4} L

(iv) 20 µL $\Rightarrow$ 20×10^{-6} L $\Rightarrow$ 2.0×10^{-5} L

(v)   0.37 µL $\Rightarrow$ 0.37×10^{-6} L $\Rightarrow$ 3.7×10^{-7} L

(vi) 0.034 L $\Rightarrow$ 0.034×1000 mL $\Rightarrow$ 34 mL

(vii) 8.4×10^{-4} L $\Rightarrow$ 0.84 mL $\Rightarrow$ 840 µL

Q2.16

Perform the following conversions:

(i)   10 mL into L   (iii) 0.067 L into mL

(ii) 11.6 µL into L   (iv) 2.6×10^{-7} L into µL

Q2.17

A solution has been prepared such that 100 mL of the solution contains 0.02 mol of sodium hydroxide (NaOH).

Calculate the concentration in moles per litre

Example 2.21

0.500 L of solution contains 4.00 g of sodium chloride, NaCl ($M_r = 58.4$).

Calculate:

(i) concentration of the solution in g L^{-1}
(ii) molar concentration in mol L^{-1}

Answers:

(i) Using [2.6], concentration $= 4.00/0.500 \Rightarrow 8.00$ g L^{-1}

(ii) Using [2.4], number of moles $n = \dfrac{m}{M_r} \Rightarrow \dfrac{4.0}{58.4} \Rightarrow 0.0685$ mol

(iii) Using [2.7], molar concentration of 0.5 L of solution $= \dfrac{0.0685}{0.500}$ mol L^{-1}

(iv) $\Rightarrow 0.137$ mol L$^{-1} \Rightarrow 0.137$ M $\Rightarrow 137$ mM

Q2.18

Calculate the concentration (in mol L^{-1}) of a solution that contains 5.6 g of sodium hydroxide, NaOH ($M_r = 40.0$), in 75 mL of solution.

Example 2.22

Calculate the mass of sodium hydroxide, NaOH ($M_r = 40$), that must be dissolved in 100 mL of solution to obtain a molar concentration of 0.50 mol L^{-1}.

Convert the volume, 100 mL, to litres: 100 mL $= 0.10$ L

Substitute in [2.7], and let n be the number of moles:

$$0.50 = \frac{n}{0.10}$$

Rearranging the equation gives: $n = 0.50 \times 0.10 \Rightarrow 0.05$ mol

Using [2.3], the required mass of 0.05 mol is equivalent to:

$$m = n \times M_r \Rightarrow 0.05 \times 40 \Rightarrow 2.0 \text{ g}$$

Q2.19

What mass of sodium chloride, NaCl ($M_r = 58.4$), when dissolved in water to give 50 mL of solution, will give a concentration of 0.10 mol L^{-1}?

Q2.20

Calculate the mass of hydrated copper sulphate, CuSO$_4$.5H$_2$O ($M_r = 249.7$), that must be dissolved into a final volume of 50 mL of solution to obtain a concentration of 0.50 M.

Other common terminologies relating to *concentration* include:

- millimoles, mmol: 1 mmol is equivalent to 1.0×10^{-3} mol
- millimolar, mM: 1 mM is equivalent to 1.0×10^{-3} M
  $= 1.0 \times 10^{-3}$ mol L^{-1}
- parts per million, ppm: 1 ppm is equivalent to 1 mg L^{-1}
- parts per billion, ppb: 1 ppb is equivalent to 1 µg L^{-1}.

Percentage concentrations are often expressed as ratios (multiplied by 100) between the masses or volumes of the solute and solvent, giving the options:

Weight of solute per volume of solution: %w/v
Volume of solute per volume of solution: %v/v
Weight of solute per weight of solution: %w/w

where the 'weights' are usually given in grams and 'volumes' in millilitres.
Note that 1.0 mL of water has a mass ('weight') of 1.0 g.

Example 2.23

Examples of typical calculations of equivalence:

- 10 mL $= 0.01$ L, 2 mL $= 0.002$ L, etc.
- 0.025 mmol in 10 mL is equivalent to $\dfrac{0.025}{0.01}$ mmol L$^{-1} = 2.5$ mM
- 0.023 mg in 10 mL is equivalent to $\dfrac{0.023}{0.01} = 2.3$ mg L$^{-1} \Rightarrow 2.3$ ppm
- 6.70×10^{-7} g in 2 mL is equivalent to $\dfrac{6.70 \times 10^{-7}}{0.002} = 0.000335$ g L$^{-1} \Rightarrow$ 335 µg L^{-1}
- 335 µg L^{-1} is equivalent to 335 ppb
- 1.2 g of solute in 50 mL of solution has a concentration of $\dfrac{1.2}{50} \times 100 = 2.4$ %w/v
- 10 mL of solute diluted to 200 mL has a concentration of $\dfrac{10}{200} \times 100 = 5$ %v/v
- 0.88 g of solute in a total of 40 g has a concentration of $\dfrac{0.88}{40} \times 100 = 2.2$ %w/w.

2.3.6 Dilutions

In most dilutions, the *amount of the solute stays the same*. In this case it is useful to use the **dilution equation**:

$$V_i \times C_i = V_f \times C_f \qquad\qquad [2.8]$$

where V_i and C_i are the initial volumes and concentrations and V_f and C_f are the final values.

The concentrations can be measured as molarities *or* mass/volume, but the *same units* must be used on each side of the equation:

$$\text{Dilution factor or ratio} \Rightarrow \frac{V_f}{V_i} = \frac{C_i}{C_f} \qquad\qquad [2.9]$$

Example 2.24

20 mL of a solution of concentration 0.3 M is transferred to a 100 mL graduated flask, and solvent is added up to the 100 mL mark. Calculate the concentration, C_f, of the final solution.

The amount of the solute is the same in the initial 20 mL as in the final 100 mL, so we can use the dilution equation [2.8]:

$$20 \times 0.3 = 100 \times C_f$$

$$C_f = \frac{20 \times 0.3}{100} \Rightarrow 0.06 \text{ M}$$

Example 2.25

It is necessary to produce 200 mL of 30 mM saline solution (sodium chloride, NaCl, in solution). Calculate the volume of a 35 g L^{-1} stock solution of saline that would be required to be made up to a final volume of 200 mL (M_r of NaCl is 58.4).

In this question, the concentrations of the two solutions are initially in different forms – moles and grams. We choose to convert 35 g L^{-1} to a molar concentration.

We know that 58.4 g (= 1 mol) of NaCl in 1.00 L gives a concentration of 1.00 M:

• 1.00 g of NaCl in 1.00 L gives a concentration of $\dfrac{1.00}{58.4}$ M

• 35.0 g of NaCl in 1.00 L gives a concentration of $35.0 \times \dfrac{1.00}{58.4} \Rightarrow 0.599$ M $\Rightarrow$ 599 mM.

We can now use [2.8] to find the initial volume, V_i, of 0.599 M (= 599 mM) saline that must be diluted to give 200 mL of 30 mM saline:

$$V_i \times 599 = 200 \times 30$$

$$V_i = \frac{200 \times 30}{599} \Rightarrow 10.02 \text{ mL}$$

Q2.21

5.0 mL of a solution of concentration 2.0 mol L^{-1} is put into a 100 mL graduated flask and pure water is added, bringing the total volume in the flask to exactly 100 mL.

Calculate the concentration of the new solution.

Q2.22

A volume, V, of a solution of concentration 0.8 mol L^{-1} is put into a 100 mL graduated flask, and pure solvent is added to bring the volume up to 100 mL. If the concentration of the final solution is 40 mM, what was the initial volume, V?

Q2.23

Calculate the volume of a 0.15 M solution of the amino acid alanine that would be needed to make up to a final volume of 100 mL in order to produce 100 mL of 30 mM alanine?

2.4   Angular Measurements

2.4.1   Introduction

In many aspects of undergraduate science, students rarely encounter the need to measure angles or solve problems involving rotations. However, angular measurements do occur routinely in a variety of practical situations. The mathematics is not difficult, and, in most cases, it is only necessary to refresh the ideas of simple trigonometry or to revisit Pythagoras!

2.4.2   Degrees and radians

There are 360° (**degrees**) in a full circle.

Example 2.26

Why are there '360' degrees in a circle?

The choice of '360' was made when 'fractions' were used in calculations far more frequently than they are now. The number '360' was particularly good because it can be divided by many different factors: 2, 3, 4, 5, 6, 8, 9, 10, 12, 15, 18, 20, 24, 30, 36, 40, 45, 60, 72, 90, 120, 180!

The **radian** is an alternative measure that is often used in calculations involving *rotations* (Figure 2.1).

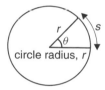

Figure 2.1   Angle in radians.

The angle, θ, in *radians* is defined as the **arc length**, s, divided by the **radius**, r, of the arc. The angle in radians is given by the simple *ratio*:

$$\theta = \frac{s}{r} \qquad [2.10]$$

$$s = r \times \theta \qquad [2.11]$$

In a *complete circle*, the *arc length*, s, will equal the *circumference* of the circle $= 2\pi r$.

Hence, the angle (360°) of a complete circle $= \dfrac{2\pi r}{r}$ radians $= 2\pi$ radians

$360° = 2\pi$ radians
$180° = \pi$ radians
$90° = \pi/2$ radians

$$1 \text{ radian} = \frac{180}{\pi} \text{ degrees} = 57.3 \ldots \text{degrees} \qquad [2.12]$$

2.4.3 Conversion between degrees and radians

$$x \text{ in radians becomes } x \times 180/\pi \text{ in degrees} \qquad [2.13]$$

$$\theta \text{ in degrees becomes } \theta \times \pi/180 \text{ in radians} \qquad [2.14]$$

In Excel, to convert an angle:

- *from* radians *to* degrees, use the function DEGREES; and
- *from* degrees *to* radians, use the function RADIANS.

Q2.24

Convert the following angles from degrees to radians or vice versa:

(i)   360° into radians   (iv) 1.0 radian into degrees
(ii) 90° into radians   (v) 2.1 radians into degrees
(iii) 170° into radians   (vi) 3.5π radians into degrees

Example 2.27

The towns of Nairobi and Singapore both lie approximately on the equator of the Earth at longitudes 36.9 °E and 103.8 °E respectively. The radius of the Earth at the equator is 6.40×10^3 km.

Calculate the distance between Nairobi and Singapore *along the surface* of the Earth.

The equator of the Earth is the circumference of a circle with radius $r = 6.40 \times 10^3$ km. Nairobi and Singapore are points on the circumference of this circle separated by an angle:

$$\theta = 103.8° - 36.9° = 66.9°.$$

Converting this angle to radians:

$$\theta = 66.9 \times \pi/180 \text{ radians} = 1.168 \text{ radians}$$

The distance on the ground between Nairobi and Singapore will be given by the arc length, s, between them. Using [2.11]:

$$s = r \times \theta = 6.40 \times 10^3 \times 1.168 \text{ km} = 7.47 \times 10^3 \text{ km}$$

Q2.25

In Figure 2.2, calculate the distance that the mass rises when the drum rotates by 40°. The radius of the drum is 10 cm.

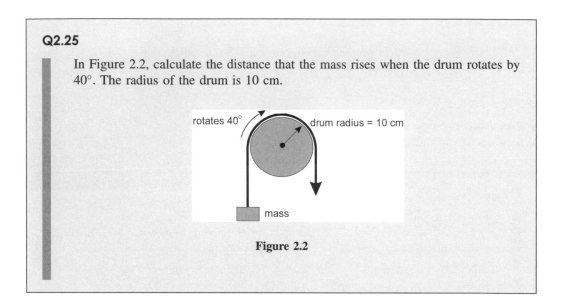

Figure 2.2

2.4.4 Trigonometric functions

In a *right-angled* triangle, the longest side is the **hypotenuse**, H.

In the triangle shown in Figure 2.3, the angle, θ, is on the left side as shown.

The side opposite the angle is called the **opposite** side, O.

The side next to the angle (but not the hypotenuse) is called the **adjacent** side, A.

The three main trigonometric functions, sine, cosine and tangent, can be calculated by taking the ratios of sides as in equation [2.15]. Many students use a simple mnemonic to remember the correct ratios: *SOHCAHTOA!*

Figure 2.3 Sides of a right-angled triangle.

$$\sin \theta = \frac{O}{H} \qquad\qquad [2.15]$$

$$\cos \theta = \frac{A}{H}$$

$$\tan \theta = \frac{O}{A}$$

Example 2.28

A car travels 100 m downhill along a road that is inclined at $15°$ to the horizontal.

Calculate the vertical distance through which the car travels.

The 100 m travelled by the car is the hypotenuse, H, of a right-angled triangle. The vertical distance to be calculated is the opposite side, O, using the angle of $\theta = 15°$:

$$\sin(15°) = \frac{O}{H} = \frac{O}{100}$$

giving:

$$O = 100 \times \sin(15°) = 100 \times 0.259 = 25.9 \text{ m}$$

Q2.26

A tree casts a shadow that is 15 m long when the Sun is at an angle of $30°$ above the horizon.

Calculate the height of the tree.

2.4.5 Pythagoras's equation

$$H^2 = O^2 + A^2 \qquad\qquad [2.16]$$

(The square on the hypotenuse of a right-angled triangle is equal to the sum of the squares on the other two sides.)

Q2.27

One side of a rectangular field is 100 m long, and the diagonal distance from one
corner to the opposite corner is 180 m. Calculate the length of the other side of
the field.

2.4.6 Small angles

When the angle θ is small (i.e. less than about 10° or less than about 0.2 radians) it is possible
to make some approximations.

In Figure 2.4:

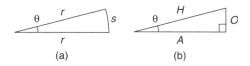

Figure 2.4 Small angles.

- The length of the arc, s, in (a) will be approximately equal to the length of the opposite
  side, O, in the right-angled triangle in (b): $O \approx s$.
- The lengths of the adjacent side, A, and the hypotenuse, H, in (b) will be approximately
  equal to the radius, r, in (a): $H \approx r$ and $A \approx r$.

If the angle θ is measured in *radians* and the *angle is small*, then:

$$\sin(\theta) = O/H \approx s/r = \theta \qquad \text{hence } \sin(\theta) \approx \theta$$

$$\tan(\theta) = O/A \approx s/r = \theta \qquad \text{hence } \tan(\theta) \approx \theta \qquad [2.17]$$

$$\cos(\theta) = A/H \approx r/r = 1 \qquad \text{hence } \cos(\theta) \approx 1.0$$

Q2.28

In the following table, use a calculator to calculate values for $\sin(\theta)$, $\cos(\theta)$ and
$\tan(\theta)$ for each of the angles listed.

Use [2.14] to calculate the angle θ in radians.

Check whether the values of $\sin(\theta)$, $\cos(\theta)$ and $\tan(\theta)$ and θ in radians agree with
[2.17].

The calculations for $\theta = 20°$ have already been performed:

θ (degrees)	$\sin(\theta)$	$\cos(\theta)$	$\tan(\theta)$	θ (radians)
20	0.3420	0.9397	0.3640	0.3491
10				
5				
1				
0				

Example 2.29

A right-angled triangle has an angle $\theta = 5°$ and an hypotenuse of length 2.0.

(i) Calculate the length of the opposite side using a trigonometric function.
We know that $\theta = 5°$ and $H = 2.0$. Using $O = H \times \sin(\theta)$:

$$O = 2.0 \times \sin(\theta) = 2.0 \times \sin(5°) = 2.0 \times 0.08716 = 0.174$$

(ii) Assume that the triangle is approximately the same as a thin segment of a circle with a radius equal to the hypotenuse, and estimate the length of the arc using a 'radian' calculation.
Converting $\theta = 5°$ into radians: $5° = 5 \times \pi/180$ radians $= 0.08727$ radians
The arc length of a circle segment with radius $r = 2.0$ is given by $s = r \times \theta$:

$$s = 2.0 \times 0.08727 = 0.175$$

The calculations for a triangle with a very small angle can often be made more easily using radians than using a trigonometric function.

Q2.29

Estimate the diameter of the Moon using the following information:

The Moon is known to be 384000 km away from the Earth, and the apparent disc of the Moon subtends an angle of about 0.57° for an observer on the Earth – as illustrated in Figure 2.5.

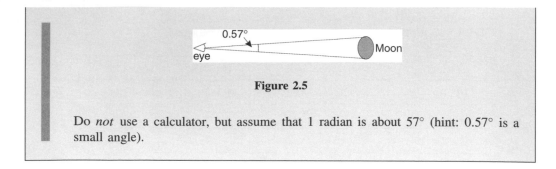

Figure 2.5

Do *not* use a calculator, but assume that 1 radian is about 57° (hint: 0.57° is a small angle).

2.4.7   Inverse trigonometric functions

The angle can be calculated from the ratios of sides by using the 'inverse' functions:

$$\theta = \sin^{-1}(O/H)$$
$$\theta = \cos^{-1}(A/H) \qquad\qquad [2.18]$$
$$\theta = \tan^{-1}(O/A)$$

Note that the above are not the reciprocals of the various functions, e.g. $\sin^{-1}(O/H)$ does *not* equal $1/[\sin(O/H)]$.

The 'inverse' function can also be written with the 'arc' prefix:

$$\theta = \arcsin(O/H)$$
$$\theta = \arccos(A/H)$$
$$\theta = \arctan(O/A)$$

Q2.30

The three sides of a right-angle triangle have lengths, 3, 4 and 5, respectively.

Calculate the value of the *smallest* angle in the triangle using:

   (i) the sine function
   (ii) the cosine function
   (iii) the tangent function

2.4.8   Calculating angular measurements

The calculation of basic angular measurements can be carried out on a calculator. Note that it is necessary to set up the 'mode' of the calculator to define whether it is using degrees (DEG) or radians (RAD).

Example 2.30 gives some examples of angle calculations on a calculator.

Example 2.30

Converting $36°$ to radians using $36 \times \pi/180$:

$$36° = 0.6283\ldots \text{ radians}$$

Converting 1.3 radians to degrees using $1.3 \times 180/\pi$:

$$1.3 \text{ radians} = 74.48\ldots°$$

Setting the calculator to DEG mode:

$$\sin(1.4) = 0.024\ldots \qquad \cos^{-1}(0.21) = 77.88\ldots°$$

Setting the calculator to RAD mode:

$$\sin(1.4) = 0.986\ldots \qquad \cos^{-1}(0.21) = 1.359\ldots \text{ radians}$$

2.4.9   Using Excel for angular measurements

When using Excel for angle calculations (see Appendix I), it is important to note that **Excel uses *radians*** as its unit of angle, *not **degrees***. To convert an angle, θ, in radians *to* degrees, use the function **DEGREES**, and to convert from degrees *to* radians, use the function **RADIANS**. Alternatively it is possible to use formulae derived from [2.13] and [2.14].

Excel uses the functions **SIN, COS** and **TAN** to calculate the basic trigonometric ratios. The inverse trigonometric functions are **ASIN, ACOS** and **ATAN**.

The value of π in Excel is obtained by entering the expression '= **PI()**'.

Example 2.31

For the following functions and formulae in Excel:

'= DEGREES(B4)' converts the angle held in cell B4 from a value given in radians to a value given in degrees.

'= B4*180/PI()' also converts the angle held in cell B4 from a value given in radians to a value given in degrees.

'= SIN(RADIANS(B4))' gives the sine of the angle (in degrees) held in cell B4.

'= DEGREES(ACOS(C3/D3))' gives the angle (in degrees) for a triangle where the length of the adjacent side is held in C3 and the length of the hypotenuse is held in D3.

Q2.31

Refer to Q2.26, where a tree casts a shadow that is 15 m long when the Sun is at an angle of $30°$ above the horizon.

Write out the formula that would be used in Excel to calculate the height of the tree, assuming that the length of the shadow was entered into cell D1 and the angle of the Sun in degrees was entered into cell D3.

3

Equations in Science

Overview

- 'How to do it' video answers for all 'Q' questions.
- Revision mathematics notes for basic mathematics:
  BODMAS, number line, fractions, powers, areas and volumes.
- Greek symbols.
- Excel tutorial: use of 'Solver'.

An equation in mathematics is a model for a *relationship* in science, and within the model it may be necessary to use letters, or other symbols, to represent the different variables of a scientific system. If we then manipulate the equation using the rules of algebra, we can generate new equations and relationships that might tell us something new about the science itself!

We start (3.1) by reviewing the basic rules and techniques of algebra, and then apply them specifically to the task of rearranging simple equations (3.2).

We then introduce (3.3) some of the more common symbols, subscripts, superscripts and other annotations that mathematics and science use to convey additional detail in equations. This then leads to further techniques (3.4) in rearranging and solving equations.

We complete the chapter by developing simple procedures for the solution of both quadratic and simultaneous equations (3.5).

3.1 Basic Techniques

3.1.1 Introduction

Equations in science frequently use letters to show how factors combine to produce a specific outcome.

As a simple example, we know that the area of a right-angled triangle is equal to half the base times the height. The formula giving the area can be written as an equation:

$$\text{Area} = {}^1\!/_2 \times \text{Base} \times \text{Height}$$

Essential Mathematics and Statistics for Science 2nd Edition   Graham Currell and Antony Dowman
Copyright © 2009 John Wiley & Sons, Ltd

In algebra we often use single letters or abbreviations to 'stand for' the various factors in an equation, and hence we could write:

$$A = 0.5 \times b \times h$$

where A = area, b = length of the base and h = height of the triangle.

In this section we introduce some basic algebraic techniques in the context of their use in handling simple equations.

Some students may find it useful to refresh their understanding of the basic mathematics associated with handling numbers, powers, fractions, etc., by visiting the Revision Mathematics section of the Website.

3.1.2 BODMAS (or BIDMAS)

The order for working out an algebraic expression follows the acronym BODMAS (or BID-MAS): Brackets first, then power Of (or Indices), followed by Divide or Multiply, and finally Add or Subtract.

Q3.1

Check that you understand the BODMAS rules, by replacing x with the number 2 and y with the number 3 in the following expressions, and calculate the values:

(i) $3x^2$ (v) $4(y - x^2)$

(ii) $(3x)^2$ (vi) $2xy^2$

(iii) $4(y - x)$ (vii) $xy - yx$

(iv) $\dfrac{x}{y^2}$ (viii) $\dfrac{x + y}{x}$

3.1.3 Algebraic equations

We can illustrate the versatility of equations by considering the changing speed of a car travelling along a road. We can use just the one equation for several situations:

$$v = u + at \qquad [3.1]$$

where v = final velocity, u = initial velocity, a = acceleration and t = time passed. (Equation [3.1] assumes that the acceleration involved is constant.)

Example 3.1

In the first situation, the car is initially travelling at $u = 10$ m s^{-1} (approx. 20 mph), and then the driver accelerates at a rate of $a = 2$ m s^{-2} (increasing by 2 m s^{-1} for each second) for a period of $t = 10$ seconds. The final velocity is given by:

Final velocity, $v = 10 + 2 \times 10 = 10 + 20 = 30$ m s^{-1} (approx. 67 mph)

Example 3.2

In another situation, a high-performance car starts from rest (with an initial velocity $u = 0$ m s^{-1}) and then accelerates at $a = 6$ m s^{-2} for $t = 4.5$ seconds, giving

Final velocity $= 0 + 6 \times 4.5 = 27$ m s^{-1} (0 to 60 mph in 4.5 seconds!)

We have used different values of v, u, a and t to describe the different situations. In this context, v, u, a and t are called the *variables* in the equation.

Q3.2

Use equation 3.1 to calculate the final velocity of a car which is initially going at 5 m s^{-1} and then accelerates at a rate of $a = 3$ m s^{-2} for a period of 4 seconds.

Q3.3

Use equation [3.1] to calculate the final velocity of a car which is initially going at 30 m s^{-1} and then accelerates at a rate of $a = -5$ m s^{-2} (the negative sign shows that this is actually a deceleration) for a period of 6 seconds.

In some types of problems, one of the factors may have the same *constant* value for a range of different situations. For example, in Einstein's well-known equation:

$$E = mc^2 \qquad [3.2]$$

E is the amount of energy (in joules) released when a mass, m (in kg), of matter is completely annihilated. E and m are both *variables* related by the *constant* c, which has a value of 3.0×10^8 m s^{-1}.

Example 3.3

Use equation [3.2] to calculate the total energy produced when 1.0 mg (1 milligram = one-thousandth of a gram) is totally annihilated in a nuclear reactor.

Now $1.0 \text{ mg} = 1.0 \times 10^{-3} \text{ g} = 1.0 \times 10^{-6} \text{ kg}$, so

$$E = 1.0 \times 10^{-6} \times (3.0 \times 10^8)^2 = 9.0 \times 10^{10} = 90\,000 \text{ MJ}$$

(enough energy to power a 1 kW heater continuously for almost 3 years).

The separate parts of an algebraic expression are often called 'terms'.
There are certain conventions which define how 'terms' in algebra should be written:

- A term with a number times a letter, e.g. $5 \times a$, would usually be written just as $5a$ (the number in a product goes before the letter, and it must *not* be written after the letter as $a5$).
- A term with '$1\times$' such as $1 \times t$ would just be written as t without the '1'.
- A product of letters, e.g. $k \times a$, could be written as ka or ak.
- Sometimes a raised *full stop* is used to represent the multiplication sign between letters: $k \times a \;(= ka)$ could also be written as $k \cdot a$ (but $3a$ would *not* be written as $3 \cdot a$).

Q3.4

Given that $v = u + at$, decide whether each of the following statements is either true, T, or false, F:

If $t = 5$, the equation could be written as $v = u + a5$	T/F
If $t = 5$, the equation could be written as $v = u + a \times 5$	T/F
If $t = 5$, the equation could be written as $v = u + 5a$	T/F
If $a = 1$, the equation could be written as $v = u + t$	T/F

Very often in science, we use letters from the Greek alphabet (see Website) in addition to those from the English alphabet. For example:

$$v = f \times \lambda$$

where v is the velocity of a wave with a frequency, f, and a wavelength, λ (lambda), and:

$$\omega = \theta/t$$

where ω (omega) is the angular velocity of a wheel when it rotates through an angle, θ (theta), in a time, t.

3.1.4 Negative values

The 'sign' of a term adds a 'sense of direction' to the value that follows it.

If, in a particular problem, we consider *upwards* to be the positive direction, then an object going upwards with a speed of 5 m s^{-1} would have a velocity of $+5$ m s^{-1}, whereas an object falling downwards with the same speed would have a velocity of -5 m s^{-1}.

For a stone thrown vertically into the air, we can still use equation [3.1] with $u =$ initial *upward* velocity and $v =$ final *upward* velocity. In this case, the acceleration will be $a = -9.8$ m s^{-2}, which is the *constant* acceleration due to gravity. It is written as negative here because we have chosen *upwards* as being the positive direction in this problem, but gravity acts *downwards*.

The equation describing the *upward* velocities becomes:

$$v = u - 9.8t \qquad\qquad [3.3]$$

Example 3.4

If a stone is thrown upwards at 10 m s^{-1}, how fast will it be travelling upwards after

(i) $t = 0.5$ seconds?
(ii) $t = 1$ second?
(iii) $t = 2$ seconds?

Using equation [3.3]:

(i) $v = 10 - 9.8 \times 0.5$ $\quad = 5.1$ m s^{-1} $\qquad$ the stone is slowing down
(ii) $v = 10 - 9.8 \times 1$ $\quad = 0.2$ m s^{-1} $\qquad$ it has almost stopped
(iii) $v = 10 - 9.8 \times 2$ $\quad = -9.6$ m s^{-1} $\qquad$ it is now falling *downwards* with increasing speed

---

There are three main rules when using signs with algebraic terms:

- The sign 'goes with' the term that follows it.
- If there is no sign, then it is assumed that the sign is *positive*.
- The order of the terms is not important as long as the 'sign' stays with its associated term.

Hence the equation $v = u - 9.8t$ can be rearranged to give $v = -9.8t + u$.

Multiplying with negative terms follows the familiar rules:

- a minus times a minus gives a plus;
- a plus times a minus gives a minus.

For example, in calculating the terms $a \times t$ and $-a \times t$:

$$
\begin{array}{llllll}
\text{If} & a = -3 \text{ and } t = -2 & \text{then} & a \times t & = & (-3) \times (-2) & = & +6 \\
\text{If} & a = -3 \text{ and } t = -2 & \text{then} & -a \times t & = & -(-3) \times (-2) & = & -6 \\
\text{If} & a = 3 \text{ and } t = -2 & \text{then} & a \times t & = & 3 \times (-2) & = & -6 \\
\text{If} & a = -3 \text{ and } t = 2 & \text{then} & -a \times t & = & -(-3) \times 2 & = & +6 \\
\end{array}
$$

3.1.5 Brackets

Brackets (also called *parentheses*) are used to *group together* the terms that are multiplied by the same *factor* (in the example below the factor is a). For example:

$$2a + 3a \Rightarrow (2 + 3)a \Rightarrow 5a$$

Similarly, *counting* 'a's in the expression $xa + ya$ gives us the equation:

$$xa + ya \Rightarrow (x + y)a$$

i.e. x lots of 'a' plus y lots of 'a' give a total of $(x + y)$ lots of 'a'.

We can also change the order of multiplication:

$$(x + y)a \Rightarrow a(x + y)$$

When *multiplying out brackets* the factor outside the bracket multiplies *every term* inside the bracket:

$$a(x + y) \Rightarrow ax + ay \Rightarrow xa + ya$$

(which equals the original expression above).

If the *multiplying factor* is a minus sign, then *every term* inside the bracket is multiplied by '-1', and changes sign:

$$-(3a - 2b) \Rightarrow -3a - (-2b) \Rightarrow -3a + 2b$$

Example 3.5

Multiply out the brackets in the following expressions:

$3(2x+3)$ $\Rightarrow 6x+9$

$v(v+w)$ $\Rightarrow v^2+vw$

$v(2x+3)$ $\rightarrow 2vx+3v$

$-3(a-b)$ $\Rightarrow -3a+(-3)\times(-b) \Rightarrow -3a+3b = 3b-3a$

$-(4p-3q)$ $\Rightarrow (-1)\times(4p-3q) \Rightarrow -4p-(-1)\times 3q \Rightarrow -4p+3q \Rightarrow 3q-4p$

$-(x-2b)$ $\Rightarrow -x+(-1)\times(-2b) \Rightarrow -x+2b \Rightarrow 2b-x$

Q3.5

Multiply out the following brackets:

(i) $3(2+x)$ (iv) $p(x+2)$

(ii) $3(2+4x)$ (v) $-3p(2-x)$

(iii) $-2(4x-3)$ (vi) $p(x+p)$

When multiplying out the product of *two* brackets, first multiply the second bracket *separately* by *every* term in the first bracket, and then multiply out *all* the different terms as in Example 3.6.

Example 3.6

To multiply out $(v-t+d)(p-t)$, multiply the second bracket by every term in the first bracket:

$$(v-t+d)(p-t) = v(p-t)-t(p-t)+d(p-t)$$

and then separately multiply out each of the three new terms in brackets:

$$= vp-vt-tp+t^2+dp-dt$$

Q3.6

Multiply out the following brackets:

(i) $(v-2)(v-2)$ [same as $(v-2)^2$]   (iv) $(v+t)(3x+y)$

(ii) $(v - t)(v + t)$ (v) $(x - 2y)(p - q)$

(iii) $(v + 4)(3 - vt)$ (vi) $(v - t + 2)(3 + v + t)$

Brackets are sometimes used in other ways. For example, the expression $\log(1 + x)$ is not a multiplication of 'log' with $(1 + x)$, but it is an *instruction* (3.3.5) to take the logarithm of $(1 + x)$.

3.1.6   Reading equations

When 'reading' a scientific equation, it is important to remember that it is a mathematical *representation* of relationships that may exist in a real-world system.

Example 3.7 shows how an equation can represent a *general* series of relationships, and how it is then possible to calculate a *specific* outcome, given a *particular* condition (i.e . the value of n in this example).

Example 3.7

The *statement* 'I think of a number, n, double it, add 6, and divide by 2 to obtain a final number, N' can be *represented* by the equation:

$$N = \frac{2n + 6}{2}$$

If the original number, n, were 4, the answer, N, would be 7, and if the original number, n, were 6, the answer, N, would be 9.

Q3.7

Write equations for N which represent the following processes, and then use them to calculate a result for the data supplied:

(i) I think of a number, n, and add 3, giving a value, N.
Use the equation to calculate N for $n = 6$.

(ii) I think of a number, n, double it, and then add 3, giving a value, N.
What is N if $n = 2$?

(iii) I think of a number, n, add 3, and then double the result to give a final value, N. What is N if $n = 2$?

(iv) I think of a number, n, add 3, double the result, take away 6, and then divide by 2, to give a final value, N.
What are the values of N if $n = 2, 6, 11$?

Example 3.8

The *statement* 'For a given amount of an ideal gas, the pressure, p, is proportional to the product of the absolute temperature, T, and the reciprocal of the volume, V' can be *represented* by the equation:

$$p \propto T \times \frac{1}{V}$$

'Product' implies multiplication in maths.

The 'reciprocal' of a value is 'one over' that value.

A 'proportional' relationship can be converted into an 'equality' equation by including a 'constant of proportionality'; in this case we include k:

$$p = k \times T \times \frac{1}{V} \Rightarrow k \times \frac{T}{V}$$

Note that for n moles of an ideal gas, $k = n \times R$, where R is the gas constant.

Q3.8

Write equations to represent the following relationships.

For example, the statement 'The area, A, of a circle can be calculated by multiplying the constant, π (Greek letter, pi), with the square of the radius, r' can be represented by the equation: $A = \pi r^2$.

(i) The area, A, of a triangle is half the base, b, times the height, h.
(ii) The distance, d, covered by a runner is equal to the product of the average speed, v, and the time, t.
(iii) The body mass index, BMI, of a person is equal to their weight (mass), m, divided by the square of their height, h.
(iv) The volume, V, of a spherical cell is four-thirds the constant, π, multiplied by the cube of the radius, r.
(v) The velocity, v, of an object that has fallen through a distance, h, is given by the square root of twice the product of h and g, where g is the acceleration due to gravity.

3.1.7 Factors

A **factor** of an expression is a term that divides *exactly* into the expression.

Example 3.9

Check your understanding of each of the following:

'3' and '10' are both factors of '30':	$30 = 3 \times 10$
'2', '5', '6' and '15' are other factors of '30':	$30 = 2 \times 15$ and $30 = 5 \times 6$
'2' and 'y' are both factors of '$2y$':	$2y = 2 \times y$
'2', '5' and 'y' are factors of '$3y + 7y$':	$3y + 7y = 10y = 2 \times 5 \times y$
$(2 + p)$ is a factor of '$(2 + p)y$'	$(2 + p)y = (2 + p) \times y$
'y' is a factor of '$(2 + p)y$'	$(2 + p)y = (2 + p) \times y$
'p' is *not* a factor of '$(2 + p)y$'	The expression cannot be rearranged to give $p \times$ (other factors)

Factorization is the process of splitting up an expression into its various factors.

Example 3.10

What are the factors of: $15xy + 12yz$?

'3' is a factor : $15xy + 12yz = 3(5xy + 4yz)$

'y' is also a factor : $3(5xy + 4yz) = 3y(5x + 4z)$

We can now also see that '$(5x + 4z)$' is another factor.

We check that the expressions are equivalent by multiplying out the brackets again, i.e. $3y(5x + 4z) = 15xy + 12yz$.

Example 3.11

What are the factors of: $2p^2 + 8p$?

'2' is a factor : $2p^2 + 8p = 2(p^2 + 4p)$

'p' is also a factor : $2(p^2 + 4p) = 2p(p + 4)$

We can now also see that '$(p + 4)$' is another factor.

We check that the expressions are equivalent by multiplying out the brackets again, i.e. $2p(p + 4) = 2p^2 + 8p$.

Q3.9

What are the factors in each of the following?

(i) $4x + 4y$
(ii) $4x + 6y$
(iii) $4x + 6x^2$
(iv) $pqx + pbx$

3.1.8 Cancelling (simplifying)

It is possible to cancel a term from both the numerator (top) and the denominator (bottom) of a fraction – as long as that term is a *factor of both the numerator and the denominator*.

Example 3.12

Check your understanding of each of the following:

(i) $\dfrac{30}{25} = \dfrac{\cancel{30}^{\,6}}{\cancel{25}_{\,5}} = \dfrac{6}{5}$ '5' is a factor of both '30' and '25' and can be cancelled from top and bottom.

(ii) $\dfrac{8}{4} = \dfrac{\cancel{8}^{\,2}}{\cancel{4}_{\,1}} = \dfrac{2}{1} = 2$ '4' is a factor of both '8' and '4'. '2' divided by '1' is just written as '2'.

(iii) $\dfrac{6x}{2y} = \dfrac{\cancel{6}^{\,3}x}{\cancel{2}_{\,1}y} = \dfrac{3x}{y}$ '2' is a factor of both '6x' and '2y' and can be cancelled from top and bottom.

(iv) $\dfrac{3x}{2x} = \dfrac{3\cancel{x}}{2\cancel{x}} = \dfrac{3}{2}$ 'x' is a factor of both '3x' and '2x' and can be cancelled from top and bottom.

(v) $\dfrac{x(4+p)}{2x} = \dfrac{\cancel{x}\,(4+p)}{2\cancel{x}} = \dfrac{(4+p)}{2}$ Note that '4' is *not* a factor of the top and cannot be cancelled with the '2' below.

(vi) $\dfrac{(4x+p)}{2x}$ cannot be simplified   The 'x' cannot be cancelled because it is not a factor of the numerator (top).

(vii) $\dfrac{3(x+p)}{x(x+p)} = \dfrac{3\cancel{(x+p)}}{x\cancel{(x+p)}} = \dfrac{3}{x}$ The bracket $(x+p)$ can be cancelled top and bottom – it is a factor of both.

Q3.10

Simplify the following equations by cancelling *where possible*:

(i) $x = \dfrac{4a}{2b}$

(iii) $x = \dfrac{y(4a + 1)}{2by}$

(ii) $x = \dfrac{4ap}{2pb}$

(iv) $x = \dfrac{2y(2a + 3)}{4by}$

3.1.9 Dividing and multiplying fractions

In the same way that numbers or variables can be *multiplied* in any order, e.g.

$$a \times b \times c = b \times c \times a = c \times a \times b \text{ etc.,}$$

the *division* of numbers or variables can also be performed in any order. In the following example a fraction is divided by a further variable:

$$\frac{\frac{a}{b}}{c} = \frac{\frac{a}{c}}{b} = \frac{a}{b \times c} \text{ etc.}$$

For example, if $a = 12, b = 3$ and $c = 4$:

$$\frac{\frac{a}{b}}{c} = \frac{\frac{12}{3}}{4} = \frac{4}{4} = 1 \text{ and } \frac{\frac{a}{c}}{b} = \frac{\frac{12}{4}}{3} = \frac{3}{3} = 1 \text{ and } \frac{a}{b \times c} = \frac{12}{3 \times 4} = \frac{12}{12} = 1$$

However, it is also useful to remember that *dividing by a fraction* is the same as *multiplying by the reciprocal of that fraction*:

$$\frac{c}{\frac{a}{b}} = c \times \frac{b}{a} \text{ and } \frac{1}{\frac{1}{b}} = 1 \times \frac{b}{1} = \frac{b}{1} = b$$

For example, if $a = 12, b = 3$ and $c = 4$:

$$\frac{4}{\frac{12}{3}} = 4 \times \frac{3}{12} = \frac{4 \times 3}{12} = \frac{12}{12} = 1 \text{ and } \frac{4}{\frac{12}{3}} = \frac{4}{4} = 1$$

Q3.11

Decide whether the answer is Yes or No for each of the following questions, whatever the values of x, y, a and b. None of x, y, a and b are zero.

(i) Does $\dfrac{x \times a}{a \times y}$ equal $\dfrac{x}{y}$?   (iv) Does $\dfrac{x/a}{y}$ equal $\dfrac{x}{y \times a}$?

(ii) Does $\dfrac{2x}{2y}$ equal $\dfrac{x}{y}$?   (v) Does $\dfrac{x}{y}$ equal $\dfrac{1}{y/x}$?

(iii) Does $\dfrac{x+2}{y+2}$ equal $\dfrac{x}{y}$?   (vi) Does $\dfrac{x/a}{y/b}$ equal $\dfrac{bx}{ay}$?

3.2 Rearranging Simple Equations

3.2.1 Rearranging the whole equation

A mathematical equation is **balanced**. The **value** on the LHS equals the **value** on the RHS. For example, in the straight line equation:

$$y = mx + c$$

the value of y is equal to the combined value of m times x then added to c.

In science, we often need to **rearrange** an equation.

For example, we may wish to make x the **subject** of the equation (i.e. on its own on the LHS of the equation), and a rearrangement will give:

$$x = \frac{y - c}{m}$$

In this unit we introduce five basic rules for performing simple rearrangement of equations.

Rule 1

An equation will continue to be **balanced** if the same mathematical operation is applied to **both** sides.

In Examples 3.13 to 3.23, we compare the rearrangement of *algebraic* equations with equations of equivalent *arithmetical* values, using the values:

$$a = 2 \qquad b = 8 \qquad c = 1$$

We start with examples of addition and division on both sides of equations. You should note that **the equations continue to balance**, i.e . both sides remain equal.

Example 3.13

Treating both sides of the equation equally:

$$a = 2c \qquad\qquad 2 = 2 \times 1$$

Adding b to *both sides*: | **Adding** 8 to *both sides*:

$$a + b = 2c + b \qquad\qquad 2 + 8 = 2 \times 1 + 8$$

Note that, in Example 3.13, the rearrangement of *arithmetic* values on the right of the page has maintained a **balanced (true) equation**. This is then also true for the rearrangement of the *algebraic* symbols on the left of the page. **The same equivalence is also true for the following examples**.

Example 3.14

Treating both sides of the equation equally:

$$b = a + 6c \qquad\qquad 8 = 2 + 6 \times 1$$

Dividing *both sides* by 2: | **Dividing** *both sides* by 2:

$$\frac{b}{2} = \frac{a + 6c}{2} \qquad\qquad \frac{8}{2} = \frac{2 + 6 \times 1}{2}$$

$$\frac{b}{2} = \frac{a}{2} + 3c \qquad\qquad 4 = 1 + 3 \times 1$$

Q3.12

Starting with $8k = 16 - 2x$

add '$2x$' to both sides of the original equation, then

subtract '$8k$' from both sides of the equation, then

divide both sides of the equation by 2.

Do you end up with $x = 8 - 4k$?

Q3.13

Starting with $\dfrac{x}{4} + 2y = 3$

subtract '$2y$' from both sides of the original equation, then multiply both sides of the equation by 4.

Do you end up with $x = 12 - 8y$?

We could use Rule 1 to explain and justify almost any rearrangement that we could perform. However, it is useful to identify four other simple rules for the more common techniques used to rearrange equations.

Rule 2

We can **'swap' the sides** (LHS and RHS) of an equation.

Example 3.15

Swapping sides:

$$b - 3a = 2c \qquad \bigg| \qquad 8 - 3 \times 2 = 2 \times 1$$

Swapping from side to side:

$$2c = b - 3a \qquad \bigg| \qquad 2 \times 1 = 8 - 3 \times 2$$

Rule 3

We can **change the *signs*** of every **term** in the equation.

This is the same as multiplying every term by '-1'.

Example 3.16

Changing the signs of every term:

$$-4c = 2a - b \qquad \bigg| \qquad -4 \times 1 = 2 \times 2 - 8 \Rightarrow -4$$

Changing the sign of every term:

$$4c = -2a + b \qquad \bigg| \qquad 4 \times 1 = -2 \times 2 + 8 \Rightarrow 4$$

Where there is a bracket, we only change the sign *in front of* the bracket, but do not change any signs *within* the bracket.

Example 3.17

Treat a bracket as a single entity:

$$b = 2(a + 3c) - 2 \qquad \bigg| \qquad 8 = 2(2 + 3 \times 1) - 2 \Rightarrow 8$$

Changing the sign of every term (*but not inside the bracket*):

$$-b = -2(a + 3c) + 2 \qquad \bigg| \qquad -8 = -2(2 + 3 \times 1) + 2 \Rightarrow -8$$

(Note that if a negative value *multiplies out* a bracket, then this will change the sign of each term inside the bracket.)

Q3.14

Change the signs on both sides of the equations to make the LHS = '$+x$':

$$(i) \quad -x = 4 - 2p$$
$$(ii) \quad -x = 3(2 - \mu) - 4p$$
$$(iii) \quad -x = 8 - (q - t)$$

3.2.2 Moving 'plus' and 'minus' terms

We can move 'plus' or 'minus' terms from one side of the equation to the other, by simply subtracting or adding (respectively) the same value to both sides of the equation:

Starting with the balanced equations:

$$b - a = 6c \qquad \Big| \qquad 8 - 2 = 6 \times 1$$

To move $-a(= -2)$ we can **add** $a(= 2)$ to both sides of the equations:

$$b - a + a = 6c + a \qquad \Big| \qquad 8 - 2 + 2 = 6 \times 1 + 2$$

and the $-a + a$ on the LHS cancels out to give:

$$b = 6c + a \qquad \Big| \qquad 8 = 6 \times 1 + 2$$

In the above equations, the 'minus' $a(= -2)$ on the LHS of the equation has the same *effect* as 'plus' $a(= +2)$ on the RHS.

Rule 4

We can **move an addition or subtraction term** from one side of the equation to the other side, but we must **change the sign** in front of the term.

Example 3.18

Moving a 'plus' or 'minus' term:

$$3a = b - 2c \qquad \Big| \qquad 3 \times 2 = 8 - 2 \times 1$$

Move the 'minus' term from the RHS to the LHS:

$$3a + 2c = b \qquad \Big| \qquad 3 \times 2 + 2 \times 1 = 8$$

When a term in brackets is moved as a unit from one side of an equation to the other, the sign *in front of* the bracket is changed, but any sign *inside* the bracket does not change.

Example 3.19

Moving a term in brackets:

$$2(b - a) + 3c = 15 \qquad \Big| \qquad 2(8 - 2) + 3 \times 1 = 15$$

Move the bracket term from the LHS to the RHS:

$$3c = 15 - 2(b - a) \qquad \begin{aligned} 3 \times 1 &= 15 - 2(8 - 2) \\ &= 15 - 2 \times 6 \end{aligned}$$

Q3.15

'Move' terms to get '$+x$' on its own on the LHS:

(i) $18 = p - x$
(ii) $3(q - p) = 10 - x$
(iii) $9 = x - p$
(iv) $(2v - t) = 8 + x - k$

3.2.3 Moving 'multiply' and 'divide' terms

We can move 'multiply' or 'divide' factors from one side of the equation to the other, by simply dividing or multiplying (respectively) both sides of the equation by the same value:

Starting with the balanced equations:

$$4a = b \qquad \Big| \qquad 4 \times 2 = 8$$

We can divide both sides by 4:

$$\frac{4a}{4} = \frac{b}{4} \qquad \Big| \qquad \frac{4 \times 2}{4} = \frac{8}{4}$$

Cancelling the '4' on the LHS gives:

$$a = \frac{b}{4} \qquad \Big| \qquad 2 = \frac{8}{4} \Rightarrow 2$$

In the above equations, the 'multiply by 4' on the LHS of the balance has the same *effect* as the 'divide by 4' on the RHS.

Rule 5

We can take a factor that **multiplies** the *whole of one side* of the equation to become the same factor **dividing** the *whole of the other side* of the equation.

Similarly we can take a factor that **divides** the *whole of one side* of the equation to become the same factor **multiplying** the *whole of the other side* of the equation.

Example 3.20

Moving a 'multiply' term:

$$5a = b + 2c \qquad \Big| \qquad 5 \times 2 = 8 + 2 \times 1$$

We can replace the '5×' on the LHS with '/5' on the RHS (note that we must divide **all of the RHS** by 5):

$$a = \frac{b + 2c}{5} \qquad \Big| \qquad 2 = \frac{8 + 2 \times 1}{5} \Rightarrow \frac{8 + 2}{5} \Rightarrow \frac{10}{5} \Rightarrow 2$$

Example 3.21

Moving a 'divide' term:

$$\frac{b}{4} + 3c = 5 \qquad \Big| \qquad \frac{8}{4} + 3 \times 1 = 2 + 3 \Rightarrow 5$$

In this case the '4' does not divide all of the LHS. We can first 'move' the '3c' (3×1) to the RHS so that the '4' divides everything on the LHS:

$$\frac{b}{4} = 5 - 3c \qquad \Big| \qquad \frac{8}{4} = 5 - 3 \times 1 \Rightarrow 5 - 3$$

Replace the '/4' on the LHS with '4×' on the RHS:

$$b = 4(5 - 3c) \qquad \Big| \qquad 8 = 4(5 - 3)$$
$$\Rightarrow 20 - 12c \qquad \Big| \qquad \Rightarrow 20 - 12$$

A term in brackets can also move across the equation as a multiplying or dividing term. Any sign *inside* the brackets does not change.

Example 3.22

Moving a divisor in brackets:

$$\frac{b}{(6c - a)} = 2 \qquad \left| \qquad \frac{8}{(6 \times 1 - 2)} = 2 \right.$$

Replace the '/()' on the LHS with '×()' on the RHS:

$$b = 2(6c - a) \qquad \left| \qquad 8 = 2(6 \times 1 - 2) \right.$$
$$\Rightarrow 12c - 2a \qquad \left| \qquad \Rightarrow 12 - 4 \right.$$

Q3.16

Multiply or divide by factors to get 'x' on its own on the LHS:

(i) $2x = 60 + 20p$

(ii) $\dfrac{x}{4} = 5 - 2q$

(iii) $x(s - m) = k$

(iv) $\dfrac{x}{(2p - q)} = 8$

3.2.4 Solving an equation for an unknown value

We often need to rearrange an equation to find the value of an unknown term.

The unknown value is often denoted by 'x'.

For example, we may need to find the value, 'x', which satisfies the following equation:

$$x + 50 = 60$$

Simple inspection of the equation will tell us that the value of 'x' must be 10.

However, we can also use a formal rearrangement of the equation, by moving the $+50$ from the LHS to become -50 on the RHS:

$$x = 60 - 50 = 10$$

The final equation giving $x = 10$ is the **solution** to the problem.

When we get the unknown (often written as 'x') on its own on the LHS, it is said that the unknown has become the **subject of the equation**.

In some situations, it is necessary to complete the rearrangement by using two or more steps.

Example 3.23

Rearrange the following equations to make 'x' the **subject** of each equation, and then **solve** for the value of 'x' by substituting for the values of $a = 2, b = 8, c = 1$.

Solve for x:	$x + 2a = b + 3c$
Move the '$2a$' term (Rule 4):	$x = b + 3c - 2a$
Substitute values:	$x = 8 + 3 \times 1 - 2 \times 2 \Rightarrow 7$
Solve for x:	$3a - x = c$
Move the '$3a$' term (Rule 4):	$-x = c - 3a$
Change all signs (Rule 3):	$x = -c + 3a$
Substitute values:	$x = -1 + 3 \times 2 \Rightarrow 5$

Q3.17

Rearrange the following equations to make 'x' the subject, and solve for 'x', by substituting with $a = 2, b = 8, c = 1$ where appropriate:

(i) $x + 4 = 5$ (iv) $3 - a = 4 - x$

(ii) $6 + x = 4$ (v) $a + x - 3 = 4$

(iii) $a + x = 6$ (vi) $25 - b - c = 30 - x$

Q3.18

Rearrange the following equations to make 'x' the subject on the LHS:

(i) $p - x = 6$ (iii) $3 - x = 0$

(ii) $3 - k = x - 4$ (iv) $3p - 4 = 4 + x + 2p$

We know that (Rule 5) multiplication or division factors can be 'moved' by dividing or multiplying (respectively) the opposite side of the equation.

Example 3.24

Make 'x' the subject of the equation: $\quad\quad\quad\quad 3x = 6 + p$

Group the RHS into a bracket first, then $\quad\quad x = \dfrac{(6 + p)}{3} \Rightarrow 2 + \dfrac{p}{3}$
divide the RHS by 3 (Rule 5):

Example 3.25

Make 'x' the subject of the equation: $\quad\quad\quad \dfrac{x}{2} = 3 + p$

Multiply the whole of the RHS by 2 (Rule 5): $\quad x = (3 + p) \times 2 \Rightarrow 2(3 + p)$
$$\Rightarrow 6 + 2p$$

We can use the same principle when we need to 'move' the unknown from the denominator of a fraction.

Example 3.26

Make 'x' the subject of the equation: $\quad 3p = \dfrac{(4q + 3)}{x}$

Multiply the LHS by x (Rule 5): $\quad\quad 3px = \dfrac{(4q + 3)}{1}$

Divide the RHS by $3p$ (Rule 5): $\quad\quad x = \dfrac{(4q + 3)}{3p}$

It is possible to divide (or multiply) both sides of the equation by a term in **brackets**.

Example 3.27

Make 'x' the subject of the equation: $\quad\quad (3 + p)x = 6 + p$

Group the RHS into a bracket first, then

divide the RHS by the term $(3 + p)$ (Rule 5): $\quad x = \dfrac{(6 + p)}{(3 + p)}$

Q3.19

Rearrange the following equations to make 'x' the subject:

(i) $5x = k - 6$

(v) $\dfrac{x}{3-p} = 5$

(ii) $mx + c = y$

(vi) $\dfrac{(2-p)x}{3-p} = 2$

(iii) $\dfrac{x}{3} + 4 = p$

(vii) $8 = \dfrac{p+q}{x}$

(iv) $(2-p)x = p + 4$

(viii) $(p-4) = \dfrac{5}{x}$

3.3 Symbols

3.3.1 Variables and constants

In science, it is common to use italic characters from both the English and Greek alphabets (see Website) to represent variables and constants in algebraic expressions.

Example 3.28

The concentration, C(mol L^{-1}), of a solution is often calculated by measuring the amount of light absorbed when passing through the solution. The relationship is given by Beer's law:

$$A = \varepsilon bC$$

- A is the *absorbance* (5.1.9) which depends on the fraction of light absorbed;
- b (cm) is the *path length* through the cell holding the solution; and
- ε (epsilon) is the *molar absorptivity*, which is a measure of the absorbance, A, for a solute concentration of 1.0 mol L^{-1} and a path length of 1.0 cm.

In this case A and C are both **variables**,

b is a **constant** for a given measurement cell, and

ε is a **constant** for the specific material (solute) being measured.

Some characters are generally reserved for specific universal constant values. For example, π (pi) $= 3.141\ldots$, and Euler's number, e $= 2.718\ldots$, represent important fundamental constants.

The symbol ∞ is used to represent 'infinity', an infinitely large number.

Q3.20

A student conducts an experiment to demonstrate Charles's law, which states that, for a given amount of an ideal gas at fixed pressure, the volume is proportional to the absolute temperature.

The relationship between p (pressure), V (volume) and T (absolute temperature) for an ideal gas is given by the equation:

$$V = \frac{nRT}{p}$$

where n is the number of moles of gas and R is the gas constant.

For each character in the above equation, identify whether it represents a variable, a constant or a fundamental constant, in this particular experiment.

3.3.2 Subscripts and superscripts

The **superscript** following the variable is reserved for the power to which that variable must be raised. For example, 'a' to the power of 3 (i.e. cubed) would be written as:

$$a^3 = a \times a \times a$$

A **subscript** following a variable is often used to identify a *specific value* for that variable. For example, the following equation can be used to describe the motion of a car at a **constant** velocity, v:

$$x_t = x_0 + vt$$

where x is the position of the car and t is the **variable** time:

- The **variable**, x_t, represents the position of the car at time, t; and
- x_0 represents the position of the car at the specific time when $t = 0$ (for a given problem x_0 is a **constant**).

In both cases, x represents the position of the car, but the subscript is used to further define the position at a specific time.

Similarly, in the following equation (see [5.24]) that describes the exponential growth of a population:

$$N_t = N_0 e^{kt}$$

- N_t is a **variable** that describes the population at the time t.
- N_0 is the specific population at the time $t = 0$. N_0 is a **constant** for a particular problem.
- e is the **fundamental exponential (Euler's) constant** ($= 2.718\ldots$), which is raised to the power of $k \times t$.
- k is a **constant** value for a particular problem which defines the rate of growth of the population (or decay if k is negative).

Subscripts can also be used to identify repeated (replicate) measurements of the same variable (e.g. 7.2.3). For example, we may use h to denote the heights of eight similar seedlings in an experiment, and it would be useful to identify each specific seedling by using a subscript after the variable. This will give us eight separate values, while still using the same variable letter, h:

$$h_1, h_2, h_3, h_4, h_5, h_6, h_7, h_8$$

Specific functions use **subscripts** to identify specific values to be used in the calculation of the function. For example, the combination function $_nC_r$ may be given values $_6C_3$, which would specifically calculate the number of ways of selecting three items from a pool of six possible items (7.5.4).

Similarly, the t-value, $t_{T,\alpha,df}$, can be read from statistical tables, for specific values of T tails, α significance and df degrees of freedom (10.1.3).

Superscripts and subscripts are also used for example in $_Z^A X$, to describe the mass number, A, and atomic number, Z, of a specific isotope of an element, X. For example, $_{17}^{37}Cl$ describes the chlorine isotope of mass number 37 and $_{17}^{35}Cl$ describes the chlorine isotope of mass number 35. However, the subscript (e.g. 17 for Cl) is often omitted as each particular element has its own unique value for Z, and its omission creates no ambiguity. So we often just write ^{37}Cl and ^{35}Cl.

3.3.3 Annotations

Additional markings can be used to **further define** specific variables. Some examples include the following:

- A value followed by an exclamation mark, '!', gives the **factorial** (7.5.2) of that value, e.g. $6! = 6 \times 5 \times 4 \times 3 \times 2 \times 1$.
- Vertical bars on either side of a number will result in only the positive value of the number, e.g. $|-6| = +6$ and $|+6| = +6$. This is called the modulus of the value.
- Square brackets are often used in chemistry to denote the 'concentration' (5.1.9) of the species identified inside the brackets, e.g. $[H^+]$ would represent the concentration of H^+ ions in a solution.

- A bar over the variable often denotes the **average** (7.2.4) of a sample of the values. For example, $\bar{h}$ (pronounced 'h-bar') may give the average of the eight values $h_1, h_2, h_3, h_4, h_5, h_6, h_7, h_8$.
- In statistics, it is common practice to use Greek symbols for variables that represent measurements on a whole population (7.2.2). For example, μ (pronounced 'mu') may be used to represent the mean (or average) height of *all* boys aged 17 in the UK, but $\bar{x}$ (pronounced 'x-bar') may represent the mean of a *sample* of boys taken from that population.
- In the special case of a logical variable, the bar over the top (7.4.5) represents the **opposite logical value**, e.g. $\overline{Yes} \Rightarrow$ NOT $Yes \Rightarrow No$.

3.3.4 Comparisons

Specific symbols are used to compare the magnitude of variables:

Symbol	Comparison
$x = y$	Value of x is equal to the value of y
$x \approx y$	Value of x is very close to the value of y
$x \sim y$	Value of x is similar to the value of y
$x \propto y$	Value of x is proportional to the value of y
$x > y$	Value of x is greater than the value of y,* e.g. $13 > 8$ and $-8 > -13$
$x < y$	Value of x is less than the value of y,* e.g. $10 < 20$ and $-20 < -10$
$x \gg y$	Value of x is *much* greater than the value of y* (typically more than 10 times greater than y)
$x \ll y$	Value of x is *much* less than the value of y* (typically less than one-tenth of y)
$x \geq y$	Value of x is greater than, *or equal to*, the value of y*
$x \leq y$	Value of x is less than, *or equal to*, the value of y*

*Note that for negative numbers 'greater than' actually means 'more positive than' and vice versa for 'less than'.

3.3.5 Functions

Some symbols are effectively **instructions** which describe a specific operation or function to be performed on the variables given.

The data used by the function 'x' in $\log(x)$, for example, is called the **argument** of the function. The symbols do not appear on their own without their argument.

Symbol	Purpose of the operation
Δx or δx	Difference between two values of x, e.g. $\Delta x = x_2 - x_1$ Δ and δ are pronounced 'delta' (Δ is 'capital' delta) (6.1.2)
Σ	Sum of a defined range of values, e.g. $\sum_1^4 h_i = h_1 + h_2 + h_3 + h_4$ Σ is pronounced 'sigma' (Σ is 'capital' sigma) (7.2.3)
$\sqrt{x}$	Square root of x
$\sqrt[3]{x}$	Cube root of x. Other roots use different numeric values
$\pm x$	Positive value and/or negative value
$\sin(x), \cos(x), \tan(x)$	Trigonometric functions (2.4.4)
$\log(x)$	Logarithm to base 10 (5.1.5)
$\ln(x)$	Logarithm to base e, 'natural logarithm' (5.1.5)
$\exp(x)$	Alternative way of writing e^x (5.1.4)

Example 3.29

Illustrating the use of various symbols

In making four repeated measurements of the pH of a solution, a student records the values 8.5, 8.6, 8.3, 8.6, and then wishes to calculate the average (or mean) value.

If we use x to represent a replicate experimental result, we could write:

$$x_1 = 8.5, x_2 = 8.6, x_3 = 8.3, x_4 = 8.6$$

To calculate the mean value, we add up all of the x_i values from x_1 to x_4:

$$\text{Sum} = \sum_1^4 x_i = x_1 + x_2 + x_3 + x_4 = 8.5 + 8.6 + 8.3 + 8.6 = 34.0$$

We then divide by the number of values, $n = 4$, for the mean value:

$$\bar{x} = 34/4 = 8.5$$

We could write this overall process as the equation:

$$\bar{x} = \frac{\sum_1^n x_i}{n}$$

Q3.21

Explain carefully what is meant by all the symbols, annotations, etc., in the following equations:

(i) $pH = -\log([H^+])$ （see 5.1.9)

(ii) $A_t = A_0 \exp(-0.693t/T_{1/2})$ （see 5.2.4)

(iii) $\chi^2 = \sum_i \frac{(O_i - E_i)^2}{E_i}$ （see 14.1.2)

3.4 Further Equations

3.4.1 Introduction

We introduced the basic rules for rearranging simple equations in 3.2:

Rule 1: Both sides of an equation must be treated *equally*.
Rule 2: The *two sides* of an equation can be swapped.
Rule 3: The signs for *every* term in an equation can be reversed (+/−).
Rule 4: A *term can be moved from one side* to the other, provided that its sign is reversed.
Rule 5: A *multiplier/divisor factor* of one side can become a *divisor/multiplier factor of the other*.

We now derive **Rule 6** for **inverse operations**, together with some additional techniques and strategies.

3.4.2 Inverse operations

Sometimes we want to 'undo' an operation and take a step backwards by using an **inverse** process. Some simple examples will illustrate the process.

If we are required to 'solve' the equation:

$$x = 5^2$$

we know that the 'square' of 5 is 25 (5 raised to the power of 2), hence we can write the solution directly:

$$x = 25$$

The **inverse process** to the 'square' is the 'square root'. Taking the square root of a squared number takes the value back to the original number! Thus:

$$\sqrt{x^2} = x$$

The following example illustrates the 'inverse' problem.

Example 3.30

Solve the following equation for 'x':

$$x^2 = 9$$

We must 'undo' the squaring process, and work 'backwards' to find a number which, when squared, gives 9.

Using Rule 1 of rearranging equations, we take the **square root of both sides** to get:

$$\sqrt{x^2} = \sqrt{9}$$

We know that $\sqrt{x^2} = x$, and finding $\sqrt{9}$ from a calculator ($= 3$), we would rewrite the above equation to give:

$$x = 3$$

In fact, the square root of 9 could be either $+3$ or -3, and we should write the solution:

$$x = \pm 3$$

We cannot tell from the mathematics which solution ($+$ or $-$) is correct, but looking back at the science of the problem will tell us whether both solutions are possible, or whether just one is correct (see Examples 3.43 and 3.44).

In the above example, the square operation on the LHS can be 'undone', and the same effect achieved, by the **inverse operation** of taking the square root of the RHS.

Example 3.31

If $x^2 = 22$, calculate the value of 'x'.

As the 'square root' is the inverse process for 'square', we take the square root of both sides of the equation:

$$\sqrt{x^2} = \sqrt{22}$$

Using a calculator to evaluate $\sqrt{22} = 4.69$, we can then write the solution:

$$x = \pm 4.69$$

We can also use the inverse operation approach to 'undo' an equation involving the reciprocal of ('one over') the unknown value.

To 'undo' a reciprocal, $1/x$, the **inverse operation** is to take the **reciprocal again** 'to get back' to 'x' (see 3.1.9):

$$\frac{1}{\frac{1}{x}} = x$$

Example 3.32

Solve the following equation for x:

$$\frac{1}{x} = 4$$

We take the reciprocal of both sides:

$$\frac{1}{\frac{1}{x}} = \frac{1}{4}$$

The LHS just equals x, which then gives:

$$x = \frac{1}{4}$$

In the above example, the reciprocal operation on the LHS can be 'undone', and the same effect achieved, by the inverse operation of taking the reciprocal of the RHS.

Rule 6

The effect of an operation on (the whole of) one side of the equation can be 'undone', and its effect replaced, by applying the inverse operation to (the whole of) the other side of the equation.

Note that the inverse operation must be applied to the whole of the other side. If necessary the other side should be enclosed within brackets before applying the inverse operation.

Performing an operation on a value, x, followed by its inverse operation will return to the original value, x. This is illustrated by the equations in the central column of Table 3.1.

Table 3.1. Inverse operations.

Operation	Inverse operations
Reciprocal of $x = 5$	Reciprocal
$\dfrac{1}{x} = 0.2$	$x = \dfrac{1}{\frac{1}{x}} = \dfrac{1}{0.2} \Rightarrow 5.0$
Square of $x = 6$	Square root
$x^2 = 36$	$x = \sqrt{x^2} = \sqrt{36} \Rightarrow \pm 6$
Square root of $x = 16$	Square
$\sqrt{x} = 4$	$x = (\sqrt{x})^2 = 4^2 \Rightarrow 16$

Inverse trigonometric functions:

Starting with $x = 30°$

$\sin(x) = 0.500$	$x = \sin^{-1}(0.500) = 30°$
$\cos(x) = 0.866$	$x = \cos^{-1}(0.866) = 30°$
$\tan(x) = 0.577$	$x = \tan^{-1}(0.577) = 30°$

Inverse of logarithmic and exponential functions (see 5.1.5):

Starting with $x = 3.6$

$\ln(x) = 1.281$	$x = e^{1.281} = 3.6$
$e^x = 36.60$	$x = \ln(36.60) = 3.6$
$\log(x) = 0.5563$	$x = 10^{0.5563} = 3.6$
$10^x = 3981$	$x = \log(3981) = 3.6$

Q3.22

Solve the following equations, finding the value of x that will make each equation true:

(i) $x^2 = 25$

(ii) $x^2 = 0.025$

(iii) $\sqrt{x} = 9$

(iv) $\log(x) = 1.3$

(v) $e^x = 26.2$

(vi) $\sin(x) = 0.34$ (give x in degrees)

(vii) $10^x = 569$

(viii) $10^x = 0.01$

3.4.3 Opening up expressions

In rearranging an equation, a given variable may often be 'buried' within several layers of mathematical operations.

We demonstrate how it is possible to remove, step by step, the layers of mathematical operations to expose the subject of the equation.

Each step in the process must follow Rule 1 – each side of the equation must be treated equally to keep the equation in 'balance'.

In Example 3.33, we start by making 'x' the subject of a simple equation.

Example 3.33

Make 'x' the subject of the following equation:

$$3x + 4 = 19$$

The 'x' is buried under two layers of operation. The first layer is the '+4' and the second layer the '×3'.

We remove the first layer by 'moving' the '4' to the RHS (using Rule 4):

$$3x = 19 - 4 \Rightarrow 15$$

We then remove the next layer by 'moving' the '×3' by dividing the RHS by 3 (using Rule 5):

$$x = \frac{15}{3} \Rightarrow 5$$

Q3.23

Rearrange the following equations, making 'x' the subject:

(i) $2x - 8 = 22$ (iii) $k = 5 - px$

(ii) $3(2 - x) = 12$ (iv) $q(2 - x) = v$

The next example demonstrates the use of the 'inverse' operation when removing layers.

Example 3.34

Make 'x' the subject of:

$$\sqrt{(8 + x^2)} = k$$

The first layer on the LHS is the square root. The effect of the square root on the LHS can be replaced by the **inverse operation of** squaring both sides of the equation (see Table 3.1):

$$8 + x^2 = k^2$$

We now move the '8' to the RHS:

$$x^2 = k^2 - 8$$

The effect of the square on the LHS can now be replaced by the **inverse operation of** taking the square root of both sides (see Table 3.1):

$$x = \sqrt{(k^2 - 8)}$$

Q3.24

Rearrange the following equations, making 'x' the subject:

(i) $x^2 + t = 25$

(ii) $(x + t)^2 = 25$

(iii) $0.6k = \sin(2x)$

(iv) $y = a - \dfrac{4}{x}$

(v) $y = e^{2x}$

(vi) $y = \sqrt{2x}$

(vii) $y = 3 \times \sqrt{(2x - 7)}$

(viii) $y = \sqrt{\left(\dfrac{2}{x} + k\right)}$

Q3.25

Solve the following equations, finding the value of 'x' that will make each equation true:

(i) $\log(x + 200) = 2.6$

(ii) $\log(2x) + 0.3 = 2.5$

(iii) $e^x = 26$

(iv) $e^{3x} = 0.87$

(v) $\log(2x + 1.2) = 0.45$

(vi) $10^{x+2} = 246$

Example 3.35 now demonstrates how this process can be expanded so that the complex exponential equation:

$$N_t = N_0 e^{-k(t - t_0)}$$

can be rearranged to make 't' the subject:

$$t = t_0 - \frac{1}{k} \times \ln\left(\frac{N_t}{N_0}\right)$$

Example 3.35

Make 't' the subject of the equation:

$$N_t = N_0 e^{-k(t-t_0)}$$

First, swap the equation round (Rule 2) to get the term containing 't' on the LHS:

$$N_0 e^{-k(t-t_0)} = N_t$$

In this example 't' is buried within several layers of mathematical operations.

In the first layer we need to move the multiplier, N_0, to the RHS, where it will become a divisor (see Rule 5):

$$e^{-k(t-t_0)} = \frac{N_t}{N_0}$$

To remove the next layer, we need to 'undo' the operation of using the exponential, 'e'. The inverse operation of 'e' is the natural logarithm, 'ln' (see Table 3.1):

$$-k(t - t_0) = \ln\left(\frac{N_t}{N_0}\right)$$

We must now change signs on both sides of the equation to make the LHS positive (Rule 3):

$$k(t - t_0) = -\ln\left(\frac{N_t}{N_0}\right)$$

We now use Rule 5 again to move the multiplier, k, to the RHS as a divisor:

$$t - t_0 = -\frac{1}{k}\ln\left(\frac{N_t}{N_0}\right)$$

Finally, we use Rule 4 to move the '$-t_0$' to the RHS to become '$+t_0$':

$$t = t_0 - \frac{1}{k}\ln\left(\frac{N_t}{N_0}\right)$$

Q3.26

(i) Make x the subject of the electrostatic force equation:

$$F = \frac{q_1 q_2}{4\pi \varepsilon_0 x^2}$$

(ii) Make θ_r the subject of the refraction equation:

$$n = \frac{\sin(\theta_i)}{\sin(\theta_r)}$$

(iii) Make a_{red} the subject of the Nernst equation:

$$E = E_0 - \frac{RT}{zF} \ln\left(\frac{a_{red}}{a_{ox}}\right)$$

(iv) Make γ the subject of the adiabatic change equation:

$$pV^\gamma = K$$

3.4.4 Grouping the unknown variable

If there is more than one term involving x, then it is necessary to **collect all the x terms** on one side of the equation.

Example 3.36

Make 'x' the subject of the equation:	$6 - 4p - 2x = 1 + 2a - 3x$
Move '6' and '$-4p$' from LHS to RHS:	$-2x = 1 + 2a - 3x - 6 + 4p$
Move '$-3x$' from RHS to LHS:	$-2x + 3x = 1 + 2a - 6 + 4p$
Simplify the LHS:	$-2x + 3x = x$
Giving:	$x = -5 + 2a + 4p \Rightarrow 2a + 4p - 5$

Q3.27

Rearrange the following equations to make 'x' the subject:

(i) $2x + 4 = x - 2$ (iv) $x = 2(p - x)$
(ii) $3 - a + x = 6 - 2x$ (v) $x(2 - a) = a(3 - x)$
(iii) $x - 3 - 2p = 5 + p - 3x$ (vi) $2(x - 1) = 3(3 - x)$

It is often necessary to **group terms together** using brackets.

Example 3.37

Using brackets to group x terms in the expression $3x + px - 2x - qx + x$, we have $3x's + px's + (-2)x's + (-q)x's +$ one x, giving:

$$3x + px - 2x - qx + x \qquad = (3 + p - 2 - q + 1)x \Rightarrow (2 + p - q)x$$

Similarly:

$$3x + ax \qquad\qquad = (3 + a)x$$
$$3x - bx + x \qquad\qquad = (3 - b + 1)x = (4 - b)x$$
$$3x - ax - 2x \qquad\qquad = (1 - a)x$$

Q3.28

Simplify each of the following expressions, by collecting the x terms together. In each expression, find the value, or expression, to write in place of the question mark, '?'.

For example, if: $2x + px = (?) \times x$

we would write $? = 2 + p$ so that $2x + px = (2 + p) \times x$.

(i) $5x - 2x = (?) \times x$ (iv) $5x - px - 2x = (?) \times x$

(ii) $3px + 2px = (?) \times x$ (v) $5x - px - 7x = (?) \times x$

(iii) $mx - 3x = (?) \times x$ (vi) $3x + px - 4x + 2kx = (?) \times x$

Example 3.38

Make 'x' the subject of the equation: $x + 4 = a - px$

Collect all x terms on the LHS: $x + px = a - 4$

Group the x's (see Example 3.37): $(1 + p)x = a - 4$

Use Rule 5 to move $(1 + p)$ to the RHS: $x = \dfrac{a - 4}{(1 + p)}$

It is important, when rearranging equations, to separate x terms from non-x terms, and to do this it is often necessary to first **expand (or open up) brackets** which contain both types of terms.

Example 3.39

Make 'x' the subject of the equation:	$3(3 + x) = 2(x + 4)$
Multiply out both brackets:	$9 + 3x = 2x + 8$
Collect all x terms on the LHS:	$3x - 2x = 8 - 9$
Giving:	$x = -1$

Q3.29

Simplify each of the following expressions, by collecting the x terms together. It will be necessary to *multiply out the brackets first* before simplifying the whole expression.

(i) $3(2 + x) - 2x + 4$

(ii) $3(2 + x) + 2(3x + 1)$

(iii) $2(3 - 3x) - 4(1 - 4x)$

(iv) $a(b - 2x) + b(2x - a)$

(v) $x(2 + a) - 3(a - x)$

(vi) $2(a - 4) - 4(a - 2 - x) + 2a - 3x$

Q3.30

Rearrange the following equations to make 'x' the subject:

(i) $2 + x = 5 - x$

(ii) $2x - a = b + 4x$

(iii) $x = 3(2 - x)$

(iv) $3x + 3 = px + 8$

(v) $3x + 3 = px + 8 - p$

(vi) $4 - qx + p = 2x - px + 9$

(vii) $x(1 - 2a) = 4a(x - 3)$

(viii) $(2 - x)p = 6$

(ix) $(2 - x)p = 6 - x$

3.4.5 Removing fractions

A fractional term in an equation can be 'cleared' by multiplying *both sides* of the equation by the **denominator** of the fraction.

Note that, provided a term (including a term in brackets) is a factor of both the top and bottom of a fraction, it can be cancelled from the fraction (3.1.8).

Example 3.40

Make 'x' the subject of the equation: $6 = \dfrac{2x}{x - 1} + \dfrac{p}{2}$

We must first remove the $(x-1)$ term from the denominator of the first fraction:

Multiply *every term* on both sides of the equation by $(x-1)$:

$$6 \times (x-1) = \frac{2x \times \cancel{(x-1)}}{\cancel{(x-1)}} + \frac{p \times (x-1)}{2}$$

Cancelling the $(x-1)$ term where possible:

$$6 \times (x-1) = 2x + \frac{p \times (x-1)}{2}$$

Multiply *every term* on both sides of the equation by '2':

$$12(x-1) = 4x + p(x-1)$$

Multiply out brackets:

$$12x - 12 = 4x + px - p$$

Collect x terms on LHS:

$$12x - 4x - px = 12 - p$$

Group the x terms with a bracket:

$$(8-p)x = 12 - p$$

Move $(8-p)$ to the RHS using Rule 5:

$$x = \frac{(12-p)}{(8-p)}$$

Q3.31

Rearrange the following equations to make 'x' the subject:

(i) $2 = \dfrac{4}{x}$

(ii) $2 = \dfrac{4-a}{x}$

(iii) $2 = \dfrac{4}{x} - a$

(iv) $\dfrac{1}{x} + \dfrac{1}{a} = 3$

(v) $\dfrac{1}{(x+1)} + \dfrac{1}{a} = 3$

(vi) $\dfrac{x}{x+1} + \dfrac{1}{a} = 3$

(vii) $\dfrac{3}{x+1} = \dfrac{4a}{x-2}$

(viii) $\dfrac{x+1}{x-1} = y$

3.5   Quadratic and Simultaneous Equations

3.5.1   Introduction

Quadratic and simultaneous equations are two further groups of equations in which we are trying to find the value of unknown variables.

3.5.2   Quadratic equations

A common form of equation where a power is involved is a 'quadratic equation'. The term **quadratic** means that the highest 'power' in the equation is 2.

For example:

$$2p^2 - p - 6 = 0 \text{ is a } quadratic\ equation \text{ in p.}$$

In general, there will be *two* possible values of p that will make a quadratic equation true, each of which will be a *possible solution*.

Example 3.41

Check, by substituting values, that both $p = 2$ and $p = -1.5$ will make the equation $2p^2 - p - 6 = 0$ true, i.e. $2p^2 - p - 6$ will equal zero.

Putting $p = 2$: $2 \times 2^2 - 2 - 6 = 8 - 2 - 6 \Rightarrow 0$ True

Putting $p = -1.5$: $2 \times (-1.5)^2 - (-1.5) - 6 = 2 \times 2.25 + 1.5 - 6$

$\Rightarrow 4.5 + 1.5 - 6 \Rightarrow 0$ True

The **general** method for finding the solutions starts by rearranging the quadratic equation into a standard form:

$$ax^2 + bx + c = 0 \qquad\qquad [3.4]$$

where a, b and c are constants in the equation, and x is the unknown variable.

With the equation in the above form, the solutions can be found by substituting the values of a, b and c into the formula:

$$x = \frac{-b \pm \sqrt{b^2 - 4ac}}{2a} \qquad\qquad [3.5]$$

The $\pm$ sign in the numerator of the equation tells us to take *two solutions*, one with the '$+$' sign and one with the '$-$' sign.

It is a straightforward process to use the above formula to solve any quadratic equation:

1. Rearrange the equation into the form $ax^2 + bx + c = 0$.
2. Work out the values in the equation equivalent to a, b and c.
3. Substitute a, b and c into the formula for the solutions.

Example 3.42

Solve the equation:	$2.4x^2 = 3.2x + 6.6$
Rearrange as in [3.4]:	$2.4x^2 - 3.2x - 6.6 = 0$
Compare directly with [3.4]:	$ax^2 + bx + c = 0$
giving the equivalences:	$a = 2.4,\ b = -3.2,\ c = -6.6$

Substitute these values in [3.5]:
$$x = \frac{-(-3.2) \pm \sqrt{(-3.2)^2 - 4 \times 2.4 \times (-6.6)}}{2 \times 2.4}$$

$$x = \frac{+3.2 \pm \sqrt{10.24 + 63.36}}{4.8} \Rightarrow \frac{+3.2 \pm \sqrt{73.6}}{4.8} \Rightarrow \frac{+3.2 \pm 8.58}{4.8}$$

giving two solutions for x: $x = 2.45$ or $x = -1.12$

Q3.32

Solve the following equations:

(i) $2x^2 - 3x + 1 = 0$ (iii) $4x = 5 - 2x^2$

(ii) $3.2x^2 - 2.5x - 0.8 = 0$ (iv) $(x - 2)^2 = 2x$

When the quadratic equation represents a scientific relationship, then the correct solution in the *scientific context* may be just one, or both, of the possible mathematical solutions. It is necessary to use other information about the scientific problem to decide which solution (or both) is correct.

Example 3.43

A cricket ball is hit directly upwards with an initial velocity of $16.0\ \mathrm{m\,s^{-1}}$. The equation for the time, t (seconds), at which it passes a height of 5.0 m is given by:

$$5.0 = 16.0t - 4.9t^2$$

Solving the quadratic equation gives *two values*, $t = 0.35$ and 2.92 s.

In the *scientific context*, the two values are the times (in seconds) at which the ball passes the height 5.0 m, first going up and then coming down!

Example 3.44

The refractive index, n, of a piece of glass is given by solving the equation $n(n - 1) = 0.75$. Find the value of n.

Expanding the equation gives $n^2 - n = 0.75$ or $n^2 - n - 0.75 = 0$.

This quadratic equation can be solved giving $n = 1.5$ or $n = -0.5$.

However, the refractive index of a material is always positive, so in this case $n = 1.5$ is the only solution.

In some equations, the value of $b^2 - 4ac$, inside the square root, may be a *negative value*. It is not normally possible to take the square root of a negative number. Hence:

$$\text{If } b^2 < 4ac \text{ the equation has } no \ real \ solutions \qquad [3.6]$$

Using *complex numbers* it is possible to derive a 'square root' from a negative number, but these techniques are beyond the mathematical scope of this book.

When the value of $b^2 - 4ac$ is *zero*, the square root term also becomes zero:

$$\text{If } b^2 = 4ac, both \text{ solutions to the equation have the } same \ value \qquad [3.7]$$

Q3.33

Solve the following equations:

(i) $x^2 - 4x + 4 = 0$ (iii) $2x^2 - 3x + 4 = 0$

(ii) $x^2 + 0.5x - 1.5 = 0$ (iv) $4x^2 + 12x + 9 = 0$

3.5.3 Simultaneous equations

The simplest example of a *simultaneous equations* problem has:

- two equations, which contain
- the same two variables, e.g. x and y, and
- with the requirement that the two equations must both be true (simultaneously) for the same values for x and y.

For example, we may have:

$$3x + 2y = 7, \text{ and}$$

$$4x + y = 6$$

Both of the above equations are true (simultaneously) if $x = 1$ and $y = 2$. Check by substituting the values into the equations and finding that they both 'balance':

$$3 \times 1 + 2 \times 2 = 7 \qquad \text{True}$$

$$4 \times 1 + 2 = 6 \qquad \text{True}$$

The solution for this pair of simultaneous equations is $x = 1$ and $y = 2$.

If the equations have three variables, e.g. x, y and z, then we will need three equations to find a solution. In general, if we have n different variables then we will need n different equations to find solutions for all n variables.

There are several different ways of finding the solution such that all the equations become true simultaneously. Analytical methods will give an exact solution, but may be mathematically difficult to perform. A graphical method can be used to find an (approximate) solution for more complicated equations.

3.5.4   Analytical solution for simultaneous equations

The most reliable analytical method for solving simultaneous equations aims to rearrange the equations so that one of the variables can be 'eliminated' from the equation.

The process follows *four* basic steps, which are illustrated by solving the simultaneous equations in Example 3.45.

Example 3.45

Find values of p and q that make both of the following equations true:

$$4p - 2q = 9 \qquad\qquad\qquad \text{[A]}$$

$$3p + 7q = 2 \qquad\qquad\qquad \text{[B]}$$

The analysis is performed in the following text:

Step 1: Rearrange the equations so that **one variable** is on its own on the LHS of both equations. For example, taking the variable p, Equation [A] becomes:

$$4p = 9 + 2q \qquad \text{[C]}$$

and Equation B becomes:

$$3p = 2 - 7q \qquad \text{[D]}$$

Step 2: The aim of this step is to arrive at an equation with only one variable involved. In this example we *eliminate* p, to get an equation with only q.
To do this, we multiply both sides of [C] by the multiplier of $p(= 3)$ in [D]:

$$3 \times 4p = 3 \times 9 + 3 \times 2q$$
$$12p = 27 + 6q \qquad \text{[E]}$$

Similarly we multiply both sides of [D] by the multiplier of p (= 4) in [C]:

$$4 \times 3p = 4 \times 2 - 4 \times 7q$$
$$12p = 8 - 28q \qquad \text{[F]}$$

From [E], we see that $27 + 6q$ equals $12p$, and from [F] we see that $8 - 28q$ also equals $12p$. So we can write this equality as a new equation:

$$27 + 6q = 8 - 28q \qquad \text{[G]}$$

Step 3: We can now solve this equation to find q, by rearranging it to make q the subject of the equation:

$$6q + 28q = 8 - 27$$
$$34q = -19$$
$$q = -0.55882$$

Note: We have 'rounded off' the value of q to 5 significant figures. It is important to be careful about rounding off too much in the middle of a problem in case a subsequent step calculates a *small difference* between *two large numbers*.

Step 4: We can now work backwards to find p, by substituting the value of q back into either [C] or [D]. Using [C]:

$$4p = 9 + 2 \times (-0.55882) \Rightarrow 7.8824$$

$$p = 7.8824/4 \Rightarrow 1.9706$$

Using the above *four* steps, the solutions (to 5 sf) to the equations are:

$$p = 1.9706 \text{ and } q = -0.55882$$

Checking the results by substituting back into the original equations, [A] and [B], to make sure that they are both *true*:

$$4 \times 1.9706 - 2 \times (-0.55882) = 9.0000$$

$$3 \times 1.9706 + 7 \times (-0.55882) = 2.0001$$

Allowing for small rounding errors, the calculated values on the RHS agree with the values in the original equations. Hence we have confirmed that the calculated values for p and q are the solutions to the original equations.

Q3.34

Two walkers, Alex and Ben, start walking towards each other along a path. Alex starts 0.5 km from a village on the path and Ben starts 3.2 km from the village. Alex walks at 1.2 m s^{-1} and Ben walks at 0.9 m s^{-1}.

We can describe the position of each man by giving the distance, d (in metres) of each man from the village as a function of time, t:

Alex: $d = 500 + 1.2 \times t$
Ben: $d = 3200 - 0.9 \times t$

Solve the simultaneous equations using an analytical method, and calculate

(i) The time taken before they meet.
(ii) Their distance from the village when they meet.

Q3.35

We can sometimes calculate the individual concentrations, C_1 and C_2, of two compounds in a chemical mixture by measuring the spectrophotometric absorbance, A_λ, of the mixture at *different* wavelengths, λ.

The results of a particular experiment produce the following simultaneous equations:

$$0.773 = 2.25 \times C_1 + 2.0 \times C_2$$

$$0.953 = 0.25 \times C_1 + 6.0 \times C_2$$

Use the analytical method to solve the equations and derive the solutions for C_1 and C_2.

It is possible to use a software procedure in Excel called 'Solver' to solve simultaneous equations – see Appendix I and the Website.

3.5.5 Graphical method for simultaneous equations

In the graphical method, the equations are plotted as lines on the same graph. The solution is given by the **co-ordinates of the point where the lines cross**.

Example 3.46

Solve the simultaneous equations:

$$y = x^2$$

$$y = 1.5 - 0.5x$$

The analysis is performed in the following text.

To solve the simultaneous equations from Example 3.46 we plot these equations as lines on a graph by choosing specific values of x and calculating the equivalent values of y for the two equations, as in the table below:

x	-2	-1	0	1	2
$y = x^2$	4	1	0	1	4
$y = 1.5 - 0.5x$	2.5	2	1.5	1	0.5

Plotting the above data gives the graph in Figure 3.1.

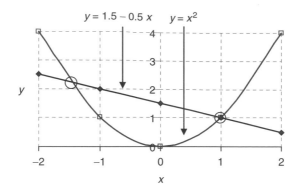

Figure 3.1   Simultaneous solutions.

   The solutions to the simultaneous equations are given by the two points where the lines cross. The intersection points are identified by two circles that have approximate values ($x = -1.45$, $y = 2.2$) and ($x = 1$, $y = 1$). These give two *approximate* solutions to the equations:

Solution 1: $x = -1.45$ and $y = 2.2$
Solution 2: $x = 1.0$ and $y = 1.0$

There will always be uncertainties in reading the co-ordinates of the crossing point from the graph. However, it is possible, once an approximate value is obtained, to redraw the graph to a much larger scale just around the crossing point.

   The particular problem in Example 3.46 can also be solved analytically by first eliminating y from the two equations, giving

$$x^2 = 1.5 - 0.5x$$

This is now a quadratic equation that can be solved using [3.5] to give two solutions for x – see Q3.36.

Q3.36

   Solve the simultaneous equations (given in Example 3.46)

$$y = x^2$$

$$y = 1.5 - 0.5x$$

   using an analytical method.

Q3.37

   Use a graphical method to solve the problem in Q3.35.

4

Linear Relationships

Overview

Website
- 'How to do it' video answers for all 'Q' questions.
- Excel tutorials: drawing $x-y$ graphs, performing linear regression, use of best-fit trendlines and calibration lines, linearization of scientific data.
- Excel files appropriate to selected 'Q' questions and Examples.

The linear relationship is one of the most common mathematical relationships used in science, and is often described on an $x-y$ graph by the equation:

$$y = mx + c$$

where m is the *slope* of the line, and c is the *intercept* of the line where it crosses the y-axis.

Note that, in various contexts, the equation of the straight line is also often written as $y = a + bx$, where a is the intercept and b is the slope. However, for consistency in this book, we will always use the form $y = mx + c$.

As a simple example of a straight line relationship, the values for temperature in the Celsius, C, and Fahrenheit, F, scales are related by the linear equation:

$$F = 1.8 \times C + 32 \qquad [4.1]$$

When plotted on an $x-y$ graph, with F on the y-axis and C on the x-axis, the above relationship appears as a straight line, as in Figure 4.1.

The slope, m, and intercept, c, of the straight line in Figure 4.1 are given by $m = 1.8$ and $c = 32$. The plotted data points show the temperatures for:

Melting ice, (0,32): $C = 0\,°\text{C}$ and $F = 32\,°\text{F}$

Essential Mathematics and Statistics for Science 2nd Edition Graham Currell and Antony Dowman
Copyright © 2009 John Wiley & Sons, Ltd

Boiling water, (100,212): $C = 100\,°C$ and $F = 212\,°F$
Point of 'equal values', $(-40, -40)$: $C = -40\,°C$ and $F = -40\,°F$

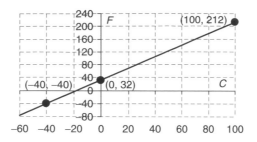

Figure 4.1   Relationship between Fahrenheit, F, and Celsius, C, temperature scales.

In Figure 4.1, the line does not pass through the origin (0, 0) of the graph.

In other scientific situations, we often find that the line does pass through the origin of the graph, i.e. the intercept, c, will be zero. In such a case, the value of one variable is *directly proportional* to the value of the other variable. For example, Beer–Lambert's law (for dilute solutions) states that the absorbance, A, of light by a solution in a spectrophotometer is directly proportional to the concentration, C, of the solution:

$$A = k \times C \qquad\qquad [4.2]$$

where $k = \varepsilon b$ is a constant for a particular measurement, ε is the absorptivity of the solute and b is the path length of the light through the solution.

There are very many situations in science where it is believed that a linear (straight line) relationship exists between two variables, but when the experimental data is plotted, the data points on the graph show random scatter away from any straight line. If the aim of the experiment is to measure the slope and/or intercept of the straight line, then the scientist has the problem of trying to interpret which straight line is the *best fit* for the given data. We see in 4.2 that the process of linear regression is a powerful mathematical procedure, which allows *information from every data point* to contribute to the identification of the (best-fit) line of regression.

Many scientific systems do not yield simple linear relationships but it is possible, in a number of specific situations, to manipulate the nonlinear data so that it can be *represented* by a linear equation – this is the process of *linearization* introduced in 4.3. The transformed data can then be analysed using the familiar process of linear regression.

As an example of linearization, a spectrophotometer measures the *transmittance*, T, of light, which is related to concentration, C, by the nonlinear equation of the form (where k is a constant):

$$T = 10^{-kC} \qquad\qquad [4.3]$$

Equation [4.3] can be 'linearized' to [4.2] by taking the logs of both sides of the equation and defining absorbance by the equation:

$$A = -\log(T) \qquad [4.4]$$

However, is important to note that the process of transforming nonlinear data to 'linear' data can distort the significance of errors in the original data. Further advice should be sought in interpreting the possible errors that might arise in the regression results from a process of linearization.

4.1 Straight Line Graph

4.1.1 Introduction

The straight line is one of the most common mathematical representations used in science.

A 'straight line', or linear, relationship occurs when the *change* in one variable (e.g. y) is *proportional* to the *change* in another variable (e.g. x), and is commonly represented by the straight line equation: $y = mx + c$.

The *slope*, m, of the line gives the *rate of change* of y with respect to x. The point at which the line crosses the y-axis is called the *intercept*, c – see Figure 4.2 below.

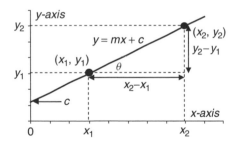

Figure 4.2   Straight line graph.

4.1.2 Plotting the graph

In an experiment, the **independent** variable is the one whose values are *chosen*, and the **dependent** variable is the one whose values are *measured*. For example, the pH (dependent variable) of the soil might be *measured* for *known* amounts (independent variable) of added lime.

An $x-y$ graph is normally plotted with x as the *independent* variable on the horizontal axis (*abscissa*) and y as the *dependent* variable on the vertical axis (*ordinate*).

We say that y *is plotted against* x, or the graph is of the form y *versus* x.

Q4.1

In an experiment to measure plant growth as a function of light exposure, different plants are exposed to different levels of light exposure, L, and the resultant growth, G, is recorded.

When the results are plotted, which of the following statements are True?

(i) G should be plotted against L (i.e. L on the x-axis).
(ii) L should be plotted against G (i.e. G on the x-axis).
(iii) It does not matter which way round the results are plotted.

4.1.3   Straight line equation

A point is located on an $x-y$ graph by using its **co-ordinates** (x, y). Note that the x-value is placed first.

Only *one* straight line can be drawn through any *two* particular points, as illustrated in Figure 4.2.

In this book we will describe the straight line by using the common equation:

$$y = mx + c \qquad\qquad [4.5]$$

where m and c are constant values that define the particular line:

- **Slope of the line** is given by the coefficient of x: m.
- **Intercept on the y-axis when $x = 0$** is given by the constant: c.

$\qquad\qquad [4.6]$

Example 4.1

Calculate the slope and intercept of the straight line described by the equation:

$$2y = 4x + 3$$

The first step is to *rearrange* the equation into the form $y = mx + c$:

Divide both sides of the equation by '2': $y = (4/2)x + (3/2)$
giving: $y = 2x + 1.5$
Comparing with the standard equation, [4.5]: $y = mx + c$
we can see that: slope $m = 2$ and intercept $c = 1.5$

Q4.2

Start by plotting the line $y = 2x + 3$ on a graph over the range $-2 < x < +2$:

(i) For each of the values of x in the table below, calculate the value of y given by the equation $y = 2x + 3$; this gives the co-ordinates (x, y) of points that will be on this straight line.

x	-2	-1	0	$+1$	$+2$
y					

(ii) Plot the co-ordinates from (i) on an $x-y$ graph, and connect with a line.
(iii) Where does the graph intercept the y-axis?
Do the following points lie on the line?
(iv) (1.5, 5.5)   Yes/No
(v) (0.5, 4)   Yes/No
(vi) $(-0.5, -1.0)$ Yes/No
Without carrying out any new calculations, sketch the following lines on the same graph as (ii):
(vii) $y = -2x + 3$
(viii) $y = 2x + 1$
(ix) $y = x + 3$

Q4.3

The length, L (m), of a simple metal pendulum as a function of the ambient temperature, T (°C), is given by the equation:

$$L = (1 + \alpha \times T) \times 1.210$$

where α is the coefficient of linear expansion of the pendulum material, and 1.210 is the length of the pendulum when $T = 0\,°C$. Assume $\alpha = 0.000\,019\,°C^{-1}$.

(i) Multiply out the bracket to obtain a straight line equation of the form $y = mx + c$.
(ii) Calculate the slope, m, and intercept, c, of the equation in (i).

Q4.4

A straight line is given by the equation:

$$3x + 4y + 2 = 0$$

Rearrange the equation to make y the subject of the equation. Hence, calculate the slope and intercept of the line.

If a particular point with co-ordinates (x, y) lies on the line, then the values of x and y will make the equation *balance*, i.e. the value of y will equal the value of $mx + c$ and the equation will be TRUE.

If the point (x, y) does *not* lie on the line, then the equation will *not balance*, and the value of y will *not* equal the value of $mx + c$.

Example 4.2

The point $(3,10)$ *lies on* the straight line $y = 2x + 4$:

Check by replacing x with '3': $2x + 4 \Rightarrow 2 \times 3 + 4 \Rightarrow 6 + 4 \Rightarrow 10$

which balances with $y = 10$ – the equation is TRUE.

The point $(4,14)$ does *not* lie on the straight line $y = 2x + 4$:

Check by replacing x with '4': $2x + 4 \Rightarrow 2 \times 4 + 4 \Rightarrow 8 + 4 \Rightarrow 12$

which does not balance with $y = 14$.

It is often necessary to calculate a value for x, given a value of y on a known straight line. We need then to rearrange equation [4.5] to make x the subject of the equation:

$$x = \frac{(y - c)}{m} \qquad [4.7]$$

Q4.5

 A straight line $(y = mx + c)$ has a slope of $+4$ and an intercept of -3. Calculate the value of x when $y = 4$

Q4.6

 The cooking time for a joint of meat is written as 40 minutes per kilogram plus 20 minutes.

(i) **Express this as an equation relating time, T (in minutes), and the mass, W (kg).**
(ii) **If there is only 2 hours available, what is the heaviest joint that could be cooked?**

Example 4.3

A car is travelling at a constant speed of 30 m s^{-1}, along a straight road *away* from a junction. I start a stopwatch with $t = 0$ when the car is 60 m away from the junction.

(i) Write down an equation that will then relate the distance, z, of the car from the junction and the time, t, in seconds on my stopwatch.
(ii) Calculate the distance of the car from the junction when $t = 3$ s.
(iii) What will be the time when the car is 210 m from the junction?

The analysis is performed in the following text.

The line and calculations for Example 4.3 can be represented on a graph of z against t:

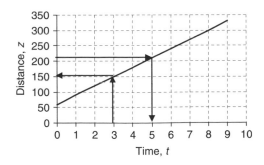

Figure 4.3 Graph for Example 4.3.

(i) The rate of change of distance, z, with time, t, is 30, and this will be a *slope*, $m = 30$ in the equation – see Figure 4.3.
The *intercept*, c, is given by the value of z when t is zero. This is given in the question as 60 m. Hence $c = 60$.
The equation is therefore:

$$z = 30 \times t + 60$$

See Figure 4.3 for the above equation plotted on a graph of z against t.

(ii) When $t = 3$, we can use the above equation to calculate:

$z = 30 \times 3 + 60 \Rightarrow 90 + 60 \Rightarrow 150$ m

(iii) To find a value for t it is necessary to rearrange the above equation into the form given by [4.7]:

$$t = \frac{(z - 60)}{30}$$

Substituting values gives: $t = (210 - 60)/30 \Rightarrow 150/30 \Rightarrow 5$ s.

Q4.7

A car is travelling at a constant speed of 20 m s^{-1} along a straight road *towards* a junction. It is at a distance $z = 140$ m away from the junction when I start my stopwatch at time $t = 0$ s.

(i) Which of the following equations will now describe the motion of the car?
  (a) $z = 20t + 140$
  (b) $z = 20t - 140$
  (c) $z = -20t + 140$
  (d) $z = -20t - 140$
(ii) What will be the time when the car passes the junction?
(iii) What will be the position of the car when $t = 20$ s?

4.1.4 Calculating slope and intercept

The **slope**, m, of the straight line (Figure 4.2) that passes through points (x_1, y_1) and (x_2, y_2) is given by the equation:

$$m = \frac{(y_2 - y_1)}{(x_2 - x_1)} \quad \text{or} \quad m = \frac{(y_1 - y_2)}{(x_1 - x_2)} \qquad [4.8]$$

The **intercept**, c, of the straight line is the value of y at the point where the line passes through the y-axis (Figure 4.2), i.e. the value of y when $x = 0$.

To calculate the *equation* of a straight line that passes through the two points (x_1, y_1) and (x_2, y_2):

1. Calculate the slope, m, using [4.8].

2. Substitute the value of m and the co-ordinates of one point (x_1, y_1) or (x_2, y_2) into [4.5] – this gives an equation where the intercept, c, is the only unknown value. Rearrange the equation to make c the subject:

$$c = y_1 - m \times x_1 \quad \text{or} \quad c = y_2 - m \times x_2$$

and calculate the value of the intercept, c.

Example 4.4

Calculate the equation of the line that passes through the points $(-2, 3)$ and $(2, 1)$.

Calculating the slope:

$$m = \frac{(y_2 - y_1)}{(x_2 - x_1)} \Rightarrow \frac{(1 - 3)}{(2 - (-2))} \Rightarrow \frac{-2}{(2 + 2)} \Rightarrow \frac{-2}{4} \Rightarrow -0.5$$

Substitute m into the equation $y_2 = mx_2 + c$, with the co-ordinates $x_2 = 2$ and $y_2 = 1$:

$$1 = (-0.5) \times 2 + c \Rightarrow -1 + c$$

Rearranging gives the intercept:

$$c = 1 + 1 = 2$$

The equation of the straight line is therefore:

$$y = -0.5x + 2$$

Q4.8

Calculate the *slopes* of the straight lines which pass through each of the following pairs of points:

(i) $(1, 1)$ and $(2, 3)$ (iv) $(2.0, 3.0)$ and $(1.0, 3.5)$

(ii) $(-1, 1)$ and $(2, 3)$ (v) $(0, 1)$ and $(1, 0)$

(iii) $(1, 1)$ and $(2, -3)$ (vi) $(2.0, 3.0)$ and $(1.0, -3.5)$

In some cases where it is necessary to calculate the coefficients of a straight line, the slope, m, may already be known. It is then possible to go directly to step 2 in the calculation process outlined above.

Q4.9

Derive the *equations* of the straight line that has:

(i) a slope of 2.0 and passes through the point (0, 3)
(ii) a slope of 0.5 and passes through the point (2, 3)
(iii) a slope of −0.5 and passes through the point (2, 3)

Q4.10

If C represents the temperature in degrees Celsius and F represents the temperature in degrees Fahrenheit, water boils at a temperature given by $C = 100$ and $F = 212$ and water freezes at a temperature given by $C = 0$ and $F = 32$. Derive a straight line equation which will give F as a function of C.

Hence calculate the value of C when $F = 0$
(hint: F and C become the y- and x-axes of a graph).

4.1.5   Intersecting lines

If two straight lines $y = m_A x + c_A$ and $y = m_B x + c_B$ meet at a point (x_0, y_0), then the values of x_0 and y_0 will make both equations true *simultaneously*:

$$y_0 = m_A x_0 + c_A$$

$$y_0 = m_B x_0 + c_B$$

Any unknown variables can (usually) be found by the methods of simultaneous equations.

Example 4.5

Calculate the *co-ordinates* of the crossing point for the two straight lines:

$$y = -0.5x + 2$$

$$y = x - 1$$

Using simultaneous equations (3.5.3) to 'solve' the above equations, we find that the values $x = 2$ and $y = 1$ will make both equations true simultaneously.

Substituting these values back into the above equations makes both equations TRUE. The point (2, 1) therefore lies on both lines. This is the point where the lines *cross*.

4.1.6 Parallel and perpendicular lines

Parallel lines have the *same* slope:

$$m_A = m_B \qquad [4.9]$$

(but *different* intercepts, $c_A \neq c_B$).

Perpendicular lines have slopes m_A and m_B given by the relationships:

$$m_A = -\frac{1}{m_B} \text{ and } m_B = -\frac{1}{m_A} \qquad [4.10]$$

Example 4.6

Calculate the equation of the line which is perpendicular to the line $y = -0.5x + 2$ and passes through the point $(-2, 3)$.

The slope of the first line, $m_A = -0.5$

Thus the slope of the *perpendicular* line, $m_B = -\left(\frac{1}{m_A}\right) \Rightarrow -\left(\frac{1}{-0.5}\right) \Rightarrow 2$

Substitute this value of m into the equation $y = m_B x + c_B$, with the co-ordinates $(-2, 3)$, and calculate the value of the intercept, c_B:

$$3 = 2 \times (-2) + c_B \Rightarrow -4 + c_B$$

Hence $c_B = 7$, and the equation of the line becomes: $y = 2x + 7$

Q4.11

A second line is *parallel* to the line $y = 3x + 2$, and passes through the point $(1,1)$. What is the equation of this second line?

Q4.12

A second line is *perpendicular* to the line $y = 3x + 2$, and passes through the point (1,1). What is the equation of this second line?

Example 4.7

Write down the equation of the line that is parallel to the x-axis (horizontal) and passes through the y-axis at $y = 2.5$.

A line that is parallel to the x-axis has slope $m = 0$.

If the line passes through the y-axis at $y = 2.5$, then $c = 2.5$.

Hence the equation of line is:

$$y = 2.5$$

Whatever the value of x, the horizontal line gives the same value, $y = 2.5$.

Q4.13

Write down the equation of the line that is parallel to the y-axis (vertical) and passes through the x-axis at $x = -1.5$.

4.1.7 Interpolation and extrapolation

These are defined as follows:

Interpolation – finding the co-ordinates of a point on the line *between* existing points.
Extrapolation – finding the co-ordinates of a point on the line *outside* existing points.

In both interpolation and extrapolation, it is necessary to calculate first the equation of the line that passes through the two points, and then find the value of x (or y) equivalent to a new value of y (or x).

Q4.14

A straight line passes through the points (2, 3) and (4, 4). Calculate the values of:

(i) y when $x = 3.8$ (iii) y where the line crosses the y-axis
(ii) x when $y = 5.6$ (iv) x where the line crosses the x-axis

Q4.15

Initially, at time $t = 0$, the height, h, of a plant is 15 cm, and it then grows linearly over a 20-day period, reaching a height of 20 cm. Derive a straight line equation which will give the height, h, as a function of the time, t, in days.

(i) What is the height when $t = 5$ days?
(ii) Estimate the height of the tree 8 days *after the end* of the 20-day period, assuming that it continues to grow at the same rate.

4.1.8   Angle of slope

The angle, θ, between the line and the x-axis (Figure 4.2) is given by the equation:

$$\tan(\theta) = \frac{(y_2 - y_1)}{(x_2 - x_1)} = m \qquad [4.11]$$

The angle, θ, can then be calculated using the inverse tangent function (2.4.7):

$$\theta = \tan^{-1}(m) \text{ or } \arctan(m) \qquad [4.12]$$

Q4.16

A map shows the height contours on the side of a hill of fairly uniform slope. A point on the 250 m contour line is seen to be 400 m horizontally from a point on the 200 m contour line.

(i) Calculate the average slope, m, between the two points.
(ii) Calculate the average slope angle of the ground with respect to the horizontal.

4.2   Linear Regression

4.2.1   Introduction

There are very many situations in science where it is believed that a linear (straight line) relationship exists between two variables. However, when experimental data is plotted, the data points on the graph show some random scatter away from any straight line. If the aim of the experiment is to measure the slope and/or intercept of the straight line, then the scientist has the problem of trying to interpret which straight line is the *best fit* for the given data.

The process of *linear regression* is a powerful mathematical procedure that calculates the position of the 'best-fit' straight line. The mathematics of the technique, also called the 'method of least squares', is covered more fully in 13.2.2.

It is fortunate that many software packages will perform the calculations necessary for linear regression, and report directly both the slope, m, and the intercept, c, of the best-fit line – see Appendix I for functions in Excel. Calculations by hand are not then required. Hence, this unit is mainly concerned with the *practicalities* of using appropriate software to perform a linear regression on scientific data.

From the point of view of experiment design, it is important to understand that the process of linear regression uses *information from every data point*. This means that the additional data points (beyond the minimum requirement of two) become *replicate* (15.1.2) data points in the calculation of the parameters of the best-fit straight line and improve the accuracy of the overall fit.

4.2.2   Linear regression

Linear regression is a mathematical process for finding the *slope*, m, and *intercept*, c, of a 'best-fit' straight line. Example 4.8 illustrates the use of Excel (see also Appendix I) to perform this process.

Example 4.8

The following $x-y$ data is entered into an Excel spreadsheet:

x	0.4	0.8	1.2	1.6	2
y	0.48	0.72	0.8	1.07	1.16

Using the Chart Wizard in Excel, it is easy to plot the graph as in Figure 4.4.

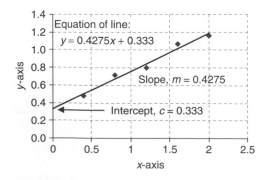

Figure 4.4   Regression of y on x.

> The best-fit straight line is drawn using the Linear Trendline option in Excel, which can also print the equation of the line on the graph.

The best-fit straight line in Figure 4.4 is called the **line of regression of y on x**.

The process of linear regression assumes that the only *significant uncertainties are in the y-direction*. The x-values for each data point are assumed to be *accurate*.

The line of regression of x on y (note reversal of x and y) will usually give a slightly different line.

Q4.17

The heart rates of athletes were measured 3 minutes after completing a series of 'step-ups' at different rates. If a process of linear regression is to be used to analyse the relationship between step-up rate and heart rate, decide which variable should be plotted on the y-axis and which variable on the x-axis.

4.2.3 Regression and correlation

It is important to know the difference (see also Chapter 13) between *regression* and *correlation*:

Linear correlation is a specific measure of the *extent* to which the values of two variables are known to change in a way that could be described by a *straight line equation*. The use of linear correlation as a statistical test is introduced in 13.1.3.

Linear regression is a process that can be used to calculate the *slope* and *intercept* of a best-fit linear equation *after* it has been shown that two variables are indeed linearly correlated.

Q4.18

In each of the following situations, one variable is plotted against another, but it is necessary to decide whether to *test for correlation* between the variables or carry out a *regression analysis*. (The process of carrying out a test for correlation is given in 13.1.3.)

(i)   Aluminium levels are recorded against sizes of fish populations to investigate whether the level of aluminium in river water affects the size of the fish population.   Correlation/ Regression

(ii)   The increase in oxygen consumption of a species of small mammal is measured as a function of body weight, in order to derive an equation that relates the two factors.   Correlation/ Regression

| (iii) | A researcher records the abundance of heather plants as a function of soil pH in different areas to see if the pH has an effect on the survival of the plant. | Correlation/ Regression |
| (iv) | Student marks are recorded for both coursework and examination to see if good students generally perform well in both and weaker students perform badly in both. | Correlation/ Regression |

4.2.4   Software calculations

Excel and other statistical software can be used to calculate both the slope and intercept of the best-fit line of regression through a set of data points – see Appendix I:

Excel Functions, f_x:

- SLOPE – calculates the best-fit slope (with no restriction on the intercept)
- INTERCEPT – calculates the best-fit intercept
- LINEST – calculates the best-fit slope (can be used to force zero intercept).

(Functions in Excel give *dynamic* results, which change if the input values change.)

Excel Data Analysis Tools:

- Regression options calculate slope and intercept (plus other data), and can force the intercept to be zero ($c = 0$).

(Data Analysis Tools in Excel are *non-dynamic*, giving a 'one-off' calculation.)

Excel Trendline:

- Calculates slope and intercept, and can force the intercept to be any chosen value.

The slope and intercept can be *displayed* on the chart but cannot be used directly in further calculations. It may be necessary to format Trendline in order to show an appropriate number of significant figures for these values.

Other statistics software packages will produce similar results. For example, Minitab produces results for slope, m, and intercept, c, in the 'equation' form:

$$C2 \; = \; c \; + \; m \; C1$$

where C1 and C2 are the columns in Minitab that hold the x- and y-data respectively.

Example 4.9

In an experiment to measure the relationship between the variables P and Q, the values of P were measured for specific values of Q, and the data entered into rows 1 and 2 of an Excel spreadsheet as below:

	A	B	C	D	E	F
1	Q	1	3	5	7	9
2	P	3.9	7.2	7.9	12.5	13.6
3	Slope =	1.235	Intercept =		2.845	

The entry '= SLOPE(B2:F2,B1:F1)' into cell B3 of the Excel spreadsheet gives the slope of the best-fit line, $m = 1.235$.
And the entry '= INTERCEPT(B2:F2,B1:F1)' into cell E3 of the Excel spreadsheet gives the intercept of the best-fit line, $c = 2.845$.
The two expressions for SLOPE and INTERCEPT assume that cells B2 to F2 hold the 'y-data' and cells B1 to F1 hold the 'x-data'.

Q4.19

Using the data from Example 4.9:

(i) Write down the equation of the line of regression, P on Q.
(ii) Plot the original data, plus the line of regression, on a graph.
(iii) Using the results from (i), calculate the value of P on the line of regression equivalent to a value of $Q = 3.4$.

Q4.20

The data in the table below gives the result of a spectrophotometric measurement, where the y variable is the absorbance, A, and the x variable is the concentration, C (mmol L^{-1}).

C	(x)	10	15	20	25
A	(y)	0.37	0.48	0.7	0.81

Use Excel and/or other statistics software to perform a linear regression to obtain values for:

(i) Slope of the best-fit straight line.
(ii) Intercept of the best-fit straight line.

4.2.5   Forcing zero intercept

Sometimes it is known that the best-fit straight line should *pass through the origin* of the graph, i.e. have zero intercept, $c = 0$.

Excel can be used to perform a linear regression while forcing a zero intercept using any of the following methods – see Appendix I:

- Function LINEST
- Data Analysis Tools > Regression
- Graph Trendline (can force the intercept to be any chosen value).

In Minitab it is possible to force a zero intercept by clicking 'Options...' in the Regression dialogue box and checking the Fit Intercept option.

Q4.21

Using the same data as in Q4.20, make the assumption that the best-fit straight line should pass through the origin of the A versus C graph.

Calculate the slope of the line of regression with zero intercept using:

(i) Excel function LINEST
(ii) Excel Data Analysis Tools
(iii) Excel Trendline
(iv) Other statistics software

4.2.6   Calibration line

The process of linear regression is frequently used to produce a best-fit **calibration line**. For example, spectrophotometric measurements often measure the absorbances, A, of solutions of known concentrations, C, and plot the best-fit straight line of A against C. The linear relationship can then be used to calculate the concentration, C_O, of the unknown solution from a measured value of its absorbance, A_O.

A line of linear regression is drawn (Figure 4.5) using n data points of known values of y, corresponding to known values of x (calibration data). The coefficients of the line of regression are calculated:

$$\text{Slope} = m$$

$$\text{Intercept} = c$$

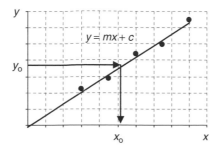

Figure 4.5 Regression of y against x.

The aim of the experiment is then to estimate the unknown value, x_0, for a given value of y_0. The 'best-estimate' x-value, x_0, of the unknown sample is then calculated by using the equation (see also [4.7]):

$$x_0 = \frac{(y_0 - c)}{m} \qquad [4.13]$$

Example 4.10

Using the data in Example 4.9 calculate the value of Q on the line of regression equivalent to a value of $P = 8.5$.

The equation of the line is:

$$P = 1.235 \times Q + 2.845$$

Rearranging this equation as in [4.13]:

$$Q = \frac{(P - c)}{m} \Rightarrow \frac{8.5 - 2.845}{1.235} \Rightarrow 4.579$$

For many calibration lines, the science of the problem is such that it is known that the calibration line must pass through the origin of the graph.

For example, in a spectrophotometric measurement it may be known that:

- the absorbance of the solution obeys Beer's Law ($A = k \times C$); and
- the spectrophotometer was accurately set up so that a solution of zero concentration records an absorbance of zero.

Then we can assume that the calibration line *must* pass through the origin of the graph of A against C.

In this case, we can force the line of regression to pass through the origin and ensure that the intercept $c = 0$.

Q4.22

Different masses w (in grams) of a sugar were dissolved in 100 mL of water and the angle of rotation θ (in degrees) of polarized light in the solution was measured in a polarimeter. The following results were obtained.

w(g)	1.5	3	4.5	6	7.5
θ (degrees)	1.3	2.7	4	5.2	6.6

(i) Assuming that pure water gives no rotation, find the best-fit equation for θ in terms of w.
(ii) Calculate the best estimate of w for a sugar solution that gives a rotation of 4.9°.

4.2.7   Interpolation and extrapolation errors

The processes of interpolation and extrapolation were introduced in 4.1.7.

There is inherent uncertainty in the values for the slope, m, and intercept, c, calculated using linear regression. Hence there will also be uncertainty in the calculation of the value, x_O, when using [4.13].

The calculation for the experimental *uncertainty* in the final value for x_O is given in 13.3.3. From this calculation it is possible to draw some general conclusions.

The calibration data should be chosen such that the unknown test value falls within the central *half* of the calibration range. For example, if the calibration data covers absorbances from 0.2 to 0.6 (not forcing through the origin), then, ideally, the absorbance of the test solution should fall within the range from 0.3 to 0.5.

Provided that the test data falls within this central calibration range, then equation [13.9] is a good estimation of the overall uncertainty in the final value of x_O.

As the interpolated test values move towards the ends of the calibration range, then the uncertainties will increase – see Figure 13.7.

If the test values fall *outside* the range of calibration data (extrapolation) then the uncertainties will begin to increase very rapidly, and great care must be taken in using extrapolated values.

Good experimental design should take the above considerations into account when planning the choice of calibration values.

4.3   Linearization

4.3.1   Introduction

We have seen (4.2) how the mathematics of linear regression is a simple, yet powerful, procedure that can be used to identify a best-fit straight line in a set of data that is expected to follow a linear relationship. However, many scientific systems do not yield simple *linear* relationships.

There are statistical methods for fitting data to specific types of curves using *nonlinear* regression but these methods are beyond the scope of this book. As an alternative to the use of nonlinear regression, it is possible, in a number of specific situations, to manipulate the nonlinear data so that it can be *represented* by a linear equation – this is the process of *linearization*. The transformed data can then be analysed using the familiar process of linear regression. However, it is important to note (4.3.4) that the process of transforming the nonlinear data can distort the significance of errors in the original data.

4.3.2   General principles

Linearization is a mathematical process whereby a nonlinear equation (e.g. involving variables P and Q) can be represented as a straight line on a suitable graph.

The original equation must be manipulated in such a way that it appears in the form:

$$\mathrm{f}(P) = m \times \mathrm{f}(Q) + c \qquad [4.14]$$

where $\mathrm{f}(P)$ and $\mathrm{f}(Q)$ are functions of the original variables, P and Q, respectively.

The above equation is in the same form as the straight line equation:

$$y = m \times x + c$$

and we can see that plotting $\mathrm{f}(P)$ against $\mathrm{f}(Q)$ (in place of y and x) will give a straight line with a slope m and an intercept c.

There are three common ways of *linearizing* a nonlinear function:

- Change of variable.
- Using logarithms (log or ln) to bring powers down onto the equation line.
- Using natural logarithms for an exponential equation.

In this chapter we only consider the 'change of variable' method.

Section 5.1.8 outlines the method for using logarithms to linearize equations with power terms of the form $T = k \times L^n$, where T and L are variables but n is an unknown constant.

Section 5.2.6 outlines the procedure for linearizing exponential equations of the form $N_t = N_0 e^{kt}$, where N_t and t are the variables.

4.3.3 Change of variable

Where possible, the simplest **linearization procedure** is to plot the functions of P and/or Q directly on the y- and x-axes of the graph.

   Example 4.11 shows that the equation that relates the area, A, of a circle to its radius, r, can be drawn as a 'straight' line by plotting the area, A, against the *square* of the radius, r^2.

Example 4.11

The following data gives the area, A, of a circle as a function of its radius, r, and as a function of r^2:

A	0	0.79	3.14	7.07	12.57	19.63	28.27
r	0	0.5	1	1.5	2	2.5	3
r^2	0	0.25	1	2.25	4	6.25	9

Figure 4.6(a) shows the nonlinear curve when we plot A directly against r. However, if we now plot A (as the y variable) against r^2 (as the x variable), we get a straight line that passes through the origin, Figure 4.6(b).

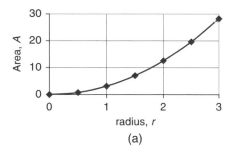

(a)

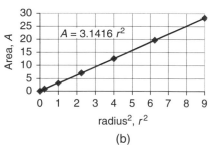

(b)

Figure 4.6

Compare the two equations:

$$A = \pi \times r^2$$

$$y = m \times x + c$$

If we plot A against r^2, we can see that the slope, m, will equal the constant π ($= 3.1416$) and that this area equation has no intercept term, giving $c = 0$:

$$A = 3.1416 \times r^2$$

Other examples of 'change of variables' are given in Table 4.1.

Table 4.1.   Examples of changes of variable for linearization.

Relationship	Equation	Variables to be plotted for linearization
Volume of a sphere, V, as a function of radius, r	$V = \dfrac{4}{3}\pi \times r^3$	Plot V against r^3 (π is a constant)
Kepler's third law of planetary motion giving period, T, against radius, r, of orbit	$T^2 = \dfrac{4\pi^2}{GM}r^3$	Plot T^2 against r^3 (π, G, M are constants)
Photoelectric effect equation of voltage, V, as a function of wavelength, λ	$V = \dfrac{hc}{e}\left(\dfrac{1}{\lambda}\right) - \phi$	Plot V against $\dfrac{1}{\lambda}$ (h, c, e, ϕ are constants)

Q4.23

The relationship between the pressure, p (Pa), the volume, V (m^3) and the absolute temperature, T (K), of 1 mole of an ideal gas is given by the equation:

$$p = RT \times \frac{1}{V}$$

where R is the gas constant.

The relationship between p and V, for constant T, is plotted on the two graphs in Figure 4.7. Figure 4.7(a) shows p against V and Figure 4.7(b) shows p against $1/V$, both plotted for constant temperatures, T.

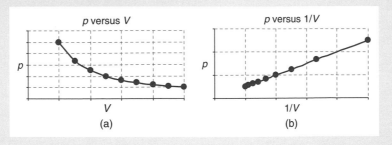

Figure 4.7

(i) Is p a linear function of V (see Figure 4.7(a))?

(ii) Is p a linear function of $1/V$ (see Figure 4.7(b))?

(iii) If p is plotted against $1/V$, what is the slope of the line in terms of R and T?

(iv) If p is plotted against $1/V$, what is the intercept of the line?

(v) In this analysis, should the line of regression of p against $1/V$ be forced to go through the origin of the graph?

(vi) If a gas at a temperature of 300 K gives a line of p against $1/V$ with a slope $m = 2500$ J mol^{-1}, calculate the value of the gas constant, R.

(vii) How would the line in Figure 4.7(b) change, if the temperature of the gas were to increase?

Q4.24

The Michaelis–Menten equation gives the initial velocity of an enzyme reaction, v, as a function of the substrate concentration, S:

$$v = \frac{v_{max} S}{K_M + S}$$

where K_M is the Michaelis–Menten constant and v_{max} would be the maximum reaction velocity for very large values of S.

Taking the reciprocal of the equation, and then dividing each term in the numerator by the denominator, $v_{max} \times S$, gives:

$$\frac{1}{v} = \frac{K_M + S}{v_{max} S} \Rightarrow \frac{K_M}{v_{max} S} + \frac{S}{v_{max} S}$$

which rearranges to:

$$\frac{1}{v} = \left(\frac{K_M}{v_{max}}\right) \times \frac{1}{S} + \frac{1}{v_{max}}$$

(i) The variables in the above equation are v and S. Compare the above equation to the straight line equation $y = mx + c$, and decide what function should be plotted on the y-axis and what function on the x-axis.

(ii) Having decided to plot the data as in (i), show how each of the constants K_M and v_{max} can be calculated from the linear regression data: slope and intercept.

4.3.4 Error warning

When using a linearization technique for nonlinear data, the transformation process will also act on the errors in the data points, and this can have the effect of distorting the importance of some of the points in the final regression process.

It is always important to be careful about interpreting the accuracy of results obtained by a linearization process, and possibly seek further guidance, particularly when relying on the accuracy of an *extrapolated* value.

5

Logarithmic and Exponential Functions

Overview

 • 'How to do it' video answers for all 'Q' questions.
• Excel tutorials: use of powers, logarithms, exponentials, linearization of exponential functions.

Logarithmic and exponential functions in mathematics are particularly useful (and common) in science because they can be used as accurate models for many real-world processes.

The mathematics of these functions can be handled very easily, provided that a few basic rules are learnt. The first unit here (5.1) concentrates on developing an understanding that will directly underpin the use of the functions in a scientific context.

The exponential function is particularly important in modelling growth or decay systems in all branches of science, e.g. elimination of a drug from the bloodstream, radioactive decay, bacterial growth, etc. Although these processes are all based on the same mathematics, the scientists in different areas have developed different ways of quantifying the processes. For example:

• '1 log' decay (or '2 log', '3 log', etc.) in drug elimination;
• 'generation time' in bacterial growth;
• 'time constant' in electronic circuits.

The second unit (5.2) demonstrates that it is easy to compare numerical calculations based on these different historical methods.

We see in 5.1.5 that a logarithm is essentially the inverse function of an exponential function. However, logarithms also have an important modelling role in their own right, particularly in systems where a *multiplication* of an input to a system produces an *additive* effect on the output, e.g. in decibel and pH scales.

The student will also often meet the use of logarithmic scales on graphs. These may be used to linearize an exponential function (5.2.6) or may simply be used to condense data which covers a wide data range.

Essential Mathematics and Statistics for Science 2nd Edition Graham Currell and Antony Dowman
Copyright © 2009 John Wiley & Sons, Ltd

5.1   Mathematics of e, ln and log

5.1.1   Introduction

The exponential constant, e, and logarithms are used extensively in science, but many students find the purpose of the exponential function and the mathematics of logarithms very confusing. However, the underlying 'rules' are quite simple and it is worthwhile spending some time getting a good understanding of the basic concepts.

5.1.2   Powers and bases

$$m^x \text{ is } \textit{pronounced 'm} \text{ to the power of } x \text{'} \qquad\qquad [5.1]$$

where m is the **base**, and x is the **exponent, power** or **index**.

In general, m can have any value, but in this unit we are mainly concerned with the two most commonly used 'bases':

- **Base '10'** – this is useful because we use a decimal system of counting.
- **Base 'e'** ($= 2.718\,28\ldots$) – this is useful because the number, e, has properties which simplify the mathematics for many problems of growth and decay in science.

The specific value, $2.718\,28\ldots$, is called Euler's number and denoted by e.

Note that, in electronic format (calculators and computers), the 'to the power of' is often expressed using '^', and sometimes by using '**', e.g. $8.0^{3.1}$ may be written as 8.0^3.1 or 8.0**3.1.

5.1.3   Properties of powers (exponents)

The properties of powers are the same for any base:

$$e^p \times e^q = e^{p+q} \qquad\qquad 10^p \times 10^q = 10^{p+q} \qquad\qquad [5.2]$$

$$\frac{e^p}{e^q} = e^p \div e^q = e^{p-q} \qquad\qquad \frac{10^p}{10^q} = 10^p \div 10^q = 10^{p-q} \qquad\qquad [5.3]$$

$$(e^p)^q = e^{p \times q} \qquad\qquad (10^p)^q = 10^{p \times q} \qquad\qquad [5.4]$$

$$e^1 = e \qquad\qquad 10^1 = 10 \qquad\qquad [5.5]$$

$$e^0 = 1 \qquad\qquad 10^0 = 1 \qquad\qquad [5.6]$$

Q5.1

Simplify each of the following (e.g. $e^2 \times e^3 = e^5$):

(i) $10^2 \times 10^3$ (v) $e^2 \times e^0$

(ii) $(10^2)^3$ (vi) $e^3 \times e^2$

(iii) $10^3 \div 10^2$ (vii) $e^5 \div e^3$

(iv) $10^3 \div 10^{-2}$ (viii) $(e^2)^3$

5.1.4 Exponential functions, e^x, 10^x

For clarity, e^x is often written as $\exp(x)$, e.g. $\exp(kT)$ can be clearer to read than e^{kT}.
Note also that the Excel function EXP(x) is used to calculate exponential value, e^x.
When using a calculator:

- e^x is calculated by using the e^x key (often the shift function of the Ln key).
- 10^x is calculated by using the 10^x key (often the shift function of the Log key).

Do not confuse the 10^x key (shift function of the Log key) with a '$\times 10^x$' key on some *calculators* which includes the multiplication to enter the *power of 10* directly in scientific notation (2.1.2), e.g. keying [1][.][6][$\times 10^x$][3] would enter the value 1.6×10^3.
   Using Excel (see also Appendix I):

- e^x is calculated by using the EXP function '=EXP(??)', and
- 10^x is calculated by typing in the formula '=10^??',

where ?? is either the value of x entered directly or the cell address where the data value of x is held.
   However, it is important to note the difference between the EXP button (power of 10) on some calculators and the EXP function (power of e) in Excel!

Q5.2

Use a calculator and/or Excel to evaluate:

(i) $10^{3.4}$ (iv) $e^{-0.42}$

(ii) $10^{-0.45}$ (v) e^1

(iii) 10^0 (vi) e^0

5.1.5   Logarithms as inverse operations of taking the power

Logarithms are *defined* as the *inverse operations* (see 3.4.2) of power terms. They have the following essential properties:

$$e^p = q \iff p = \ln(q) \qquad\qquad 10^p = q \iff p = \log(q) \qquad [5.7]$$

For example: if $e^x = 8.0$ then $x = \ln(8.0)$ $\Rightarrow 2.08$ (3 sf)

and the inverse: if $\ln(y) = 1.7$ then $y = e^{1.7}$ $\Rightarrow 5.47$ (3 sf)

$\ln(q)$ is the usual way of writing $\log_e(q)$, the logarithm of q to base e

$\log(q)$ is the usual way of writing $\log_{10}(q)$, the logarithm of q to base 10.

On a calculator, e^x and 10^x usually appear as the *second* (or *shift*) *functions* on the keys for ln and log respectively.

The function $\log_e(x)$, or $\ln(x)$, is also called the **natural**, or Naperian, logarithm.

The values for $\log(x)$ and $\ln(x)$ for any *positive* value of x can be found directly by using the log and ln keys on a calculator, or by using the functions LOG and LN in Excel (see also Appendix I). It is not possible to calculate the logarithm of a *negative* number.

Q5.3

Using a calculator or Excel, evaluate:

(i)   log(348)   (vi)   log(0.5)
(ii) log(34.8)   (vii) log(2)
(iii) log(3.48)   (viii) log(20)
(iv) log(100)   (ix) $\ln(e^1)$
(v)   log(0.01)   (x)   ln(10)

We can use the inverse operations performed by logarithms to solve some simple equations where the unknown value is in a power term.

Example 5.1

Use an inverse operation to solve equations of the form: $e^{2x} = 3.7$.

Comparing the above equation with $e^p = q$:

$$p = 2x \text{ and } q = 3.7$$

Using the inverse operation from [5.7]:

$$e^p = q \Leftrightarrow p = \ln(q)$$

we get that:

$$2x - \ln(3.7) \Rightarrow 1.308$$

We can divide both sides by 2 to calculate:

$$x = 1.308/2 \Rightarrow 0.654 (3 \text{ sf})$$

Example 5.2

Solve an equation of the form: $\log(0.43x) = -0.067$.

Comparing the above equation with $\log(q) = p$:

$$q = 0.43x \text{ and } p = -0.067$$

Using the inverse operation [5.7] we can see that if $\log(q) = p$ then $q = 10^p$ and we get that:

$$0.43x = 10^{-0.067} \Rightarrow 0.857$$

We can divide both sides by 0.43 to calculate:

$$x = 0.857/0.43 \Rightarrow 1.99 (3 \text{ sf})$$

Q5.4

Calculate x in the following by using inverse operations from [5.7]:

(i) $e^x = 22$

(ii) $e^{3x} = 12$

(iii) $e^{-2x} = 3.1 \times 10^{-3}$

(iv) $\ln(x) = 1.68$

(v) $\ln(-0.81x) = 0.37$

(vi) $10^x = 18$

(vii) $10^{2x} = 18$

(viii) $10^{-x} = 1.0 \times 10^{-7}$

(ix) $\log(x) = -9.3$

(x) $\log(-2.6x) = 3.2$

5.1.6 Important properties of logarithms

As the logarithm is the inverse operation of the power, taking the logarithm of a power can bring the exponent value down onto the equation line and simplify the expression.

We have two very important sets of relationships, which are very useful when using logarithms to simplify 'power' equations:

$$\ln(e^p) = p \qquad \log(10^p) = p \qquad\qquad [5.8]$$

and:

$$\ln(p^q) = q \times \ln(p) \qquad \log(p^q) = q \times \log(p) \qquad\qquad [5.9]$$

Other useful values and properties of logarithms include the following relationships:

$$\ln(pq) = \ln(p) + \ln(q) \qquad \log(pq) = \log(p) + \log(q) \qquad\qquad [5.10]$$

$$\ln\left(\frac{p}{q}\right) = \ln(p) - \ln(q) \qquad \log\left(\frac{p}{q}\right) = \log(p) - \log(q) \qquad\qquad [5.11]$$

$$\ln(e) = 1 \qquad\qquad \log(10) = 1 \qquad\qquad [5.12]$$
$$\ln(1) = 0 \qquad\qquad \log(1) = 0 \qquad\qquad [5.13]$$
$$\ln(0) = -\infty \qquad\qquad \log(0) = -\infty \qquad\qquad [5.14]$$

$$\log(2) \approx 0.30 \qquad\qquad [5.15]$$

$$\ln(x) = 2.30 \times \log(x) \qquad\qquad [5.16]$$

This equation is true (to 3sf) for any value of x.

Example 5.3

Check that you understand the following calculations (to 3 sf where appropriate).

Using equation [5.8]:

$$\ln(e^{kT}) = kT$$

$$\log(1000) = \log(10^3) = 3$$

$$\log(0.01) = \log(10^{-2}) = -2$$

Using equation [5.9]:

$$\ln(8^3) = 3 \times \ln(8) = 6.24$$

$$\log(e^{kT}) = kT \times \log(e) = 0.434kT$$

Using equation [5.10]:

$$\ln(3 \times 8) = \ln(3) + \ln(8) = 1.099 + 2.079 = 3.18$$

$$\log(200) = \log(2 \times 100) = \log(2) + \log(100) = 0.301 + 2 = 2.30$$

Using equation [5.11]:

$$\ln(3/8) = \ln(3) - \ln(8) = 1.099 - 2.079 = -0.98$$

$$\log(0.02) = \log(2/100) = \log(2) - \log(100) = 0.301 - 2 = -1.70$$

Q5.5

Use equations [5.8] to [5.16] to evaluate the following expressions *without* using a calculator:

(i) $\log(10^{-0.3})$ Hint: use equation [5.8]

(ii) $\ln(e^{0.62})$ Hint: use equation [5.8]

(iii) $\log(2)$ Hint: use equation [5.15]

(iv) $\log(20)$ Hint: $\log(20) = \log(2 \times 10)$ and use equation [5.10]

(v) $\log(0.5)$ Hint: $\log(0.5) = \log(1 \div 2)$ and use equation [5.11]

(vi) $\ln(1000)$ Hint: use equation [5.16]

(vii) $\ln(2)$ Hint: use equations [5.16] and [5.15]

(viii) $\log(2^{3.1})$ Hint: use equations [5.9] and [5.15]

Example 5.4

Show that, by taking natural logarithms of both sides, the equation:

$$N_t = N_0 e^{kt}$$

can be converted to the form:

$$\ln(N_t) = \ln(N_0) + kt$$

As the base in the equation is e, we will take *natural* logs of both sides:

$$\ln(N_t) = \ln(N_0 \times e^{kt})$$

then using equation [5.10] for the logarithm of a product:

$$\ln(N_t) = \ln(N_0) + \ln(e^{kt})$$

and using equation [5.8]:

$$\ln(N_t) = \ln(N_0) + kt$$

5.1.7 Solving 'power' equations with logarithms

In solving equations, we need to make the unknown value (e.g. x) the subject of the equation. When the variable x is in a power term, we need to bring it down onto the equation line by using [5.7], [5.8] or [5.9].

Examples 5.1 and 5.2 show how we can use equation [5.7] with simple equations involving powers of 10 and e.

With more complex equations we often need to make some initial rearrangements before taking the inverse operation.

Example 5.5

Solve the following equation to find t: $N_t = N_0 e^{kt}$

where $N_0 = 4260$, $N_t = 520$ and $k = -0.048$ day^{-1}. As the units of k are 'per day' we know that the units of time, t, will be 'days'.

Using the rules for rearranging equations developed in 3.2 and 3.4:

Swap the equation round to put the t term on the LHS (Rule 2), and substituting values:

$$4260 \times e^{-0.048 \times t} = 520$$

Divide both sides by 4260 to leave the 'e term' clear on the LHS (Rule 5):

$$e^{-0.048 \times t} = 520/4260 \Rightarrow 0.1221 (4 \text{ sf})$$

Now we can use the inverse operation for e, Rule 6 (3.4.2), or equation [5.7]:

$$-0.048 \times t = \ln(0.1221) \Rightarrow -2.103$$

and finally divide both sides by -0.048 (Rule 5):

$$t = -2.103/(-0.048) \Rightarrow 43.8 \text{ days } (3 \text{ sf})$$

The calculation shows that the exponential decay will fall from an initial value, $N_0 = 4260$, to a value $N_t = 520$, after a time $t = 43.8$ days.

In problems where the base of the power is neither 10 nor e, it is necessary to take logarithms of both sides of the equation and then to use equation [5.9].

Example 5.6

To solve equations of the form $8^{6y} = 0.6$:

The 'base' is neither e nor 10 so we can take either ln (base e) or log (base 10) of both sides.
For example, taking logs (base 10) of both sides gives:

$$\log(8^{6y}) = \log(0.6)$$

Using equation [5.9] we can write the LHS of the above equation as:

$$\log(8^{6y}) = 6y \times \log(8) \Rightarrow 6y \times 0.903$$

Hence combining the above two equations:

$$6y \times 0.903 = \log(0.6) \Rightarrow -0.2218$$

giving:

$$y = -0.2218/(6 \times 0.903) \Rightarrow -0.0409$$

Q5.6

Solve each of the following equations (i.e. find the value of p):

(i) $2e^{3p} = 22$ (iii) $4^p = 33$

(ii) $1.2 \times 10^{2p} = 18$ (iv) $0.3 = 0.2^p$

5.1.8 Using logarithms to linearize 'power' equations

It is sometimes possible to use linear regression for problems involving power equations by following these steps:

1. Take logarithms of both sides to bring all the powers onto the equation line.
2. Choose appropriate variables (4.3.3) to plot the equation as a straight line.
3. Perform a linear regression to calculate 'best-fit' slope and intercept.
4. Interpret the unknown values from the slope and intercept.

Example 5.7

A student believes that the period of swing, T, of a simple pendulum is related to its length, L, by a power equation of the form:

$$T = k \times L^n$$

where k is an unknown constant and n is an unknown constant power.

The student records the following periods, T, for different pendulum lengths, L.

L(m)	0.20	0.40	0.60	0.80	1.00
T(s)	0.89	1.25	1.65	1.71	1.98

The calculation for the best estimates for the values of k and n is given in the following text.

The data for T and L recorded in Example 5.7 does not give a straight line, so the first step is to linearize the equation $T = k \times L^n$, by taking logarithms of both sides:

$$\log(T) = \log(k \times L^n)$$

Expanding the RHS using equation [5.10]:

$$\log(T) = \log(k \times L^n) \Rightarrow \log(k) + \log(L^n)$$

Using equation [5.9] to bring out the power, n, from $\log(L^n)$:

$$\log(T) = \log(k) + n \times \log(L)$$

If we compare this equation with:

$$y = c + m \times x$$

we can see that $\log(T)$ is equivalent to y and $\log(L)$ is equivalent to x.
Plotting $\log(T)$ on the y-axis against $\log(L)$ on the x-axis:

$\log(L)$	−0.6990	−0.3979	−0.2219	−0.0970	0.0000
$\log(T)$	−0.0506	0.0969	0.2175	0.2330	0.2967

should give a straight line (4.3.2) with a slope $m = n$ and an intercept $c = \log(k)$.
   The calculation for Example 5.7 is performed on the Website using Excel, and a linear regression analysis gives values for:

$$\text{Slope,} \quad m = 0.4944$$

$$\text{Intercept,} \quad c = 0.2987$$

From the *slope* we calculate n directly: $n(= m) = 0.4944$
From the *intercept* we know that $\log(k)(= c) = 0.2987$
Taking the inverse of the log we calculate that:

$$k = 10^{0.2987} \Rightarrow 1.989$$

This gives a best-estimate equation:

$$T = 1.989 \times L^{0.4944}$$

In fact the true equation has values (g is the acceleration due to gravity):

$$T = 2\pi\sqrt{\frac{L}{g}} \Rightarrow 2.006 \times L^{0.5}$$

Note that it would be possible to use *natural* logarithms for the above calculation. In this case the value of n would still equal the slope, $n = m$, but the value of k would be calculated using $k = e^c$.

Q5.7

For each of the following equations, identify the functions that would be used for each axis to plot the equation as a straight line. In each case, identify how the unknown power can be determined from the regression line.

(i) $V = pA^k$ V is the volume of an animal of surface area A (p and k are unknown constants)

(ii) $E = \sigma(T+273)^z$ E is the intensity of radiation emitted from an object at a temperature $T\,°C$ (z and σ are unknown constants)

5.1.9 Logarithmic scales

Many systems in science are measured as a **ratio** of values. For example, in microbiology a '1-log' *decrease* in bacterial population is used to indicate a drop in population by a *factor* of 10 to one-tenth of the initial value. Similarly a '2-log' decrease would be a drop in population to one-hundredth of the initial value etc.

Q5.8

A bacterial population has an initial value of 5.0×10^7. Calculate the population following:

(i) a '1-log' decrease
(ii) a '3-log' decrease

The *decibel scale* of loudness, L, appropriate to the human ear, compares the power density, P, of the sound with the power density, $P_0 = 1.0 \times 10^{-12}$ W m^{-2}, which is the quietest sound that can just be heard in normal hearing:

$$\text{Loudness, } L(\text{dB}) = 10 \log(P/P_0) \qquad\qquad [5.17]$$

When comparing two sounds the *difference in loudness*, $L_1 - L_2$, will be given by:

$$L_1 - L_2 = 10 \log(P_1/P_2) \text{ dB} \qquad\qquad [5.18]$$

The *difference* in loudness depends on the *ratio* of the power densities, P_1 and P_2. As P_1 and P_2 have the same units, the ratio has no units, and $10 \log(P_1/P_2)$ is simply a number.

Example 5.8

If the power density of sound is doubled, $P_1/P_2 = 2$, calculate the increase in loudness.

Difference in loudness: $L_1 - L_2 = 10 \log(2) = 10 \times 0.30 = 3.0$ dB

A *doubling* of power gives an *addition* of 3 dB.

Q5.9

A sound has an initial loudness of 70 dB. Calculate the new loudness using [5.18] if the power density of the sound is:

(i) doubled (i.e. by a factor of 2)	(iv) halved
(ii) increased by a factor of 8	(v) reduced by a factor of 8
(iii) increased by a factor of 100	(vi) reduced by a factor of 100

Another important area where students will encounter the use of logarithms is in electro-chemistry.

The *acidity* of a solution depends on the *concentration* of hydrogen ions $[H^+]$ expressed in units of mol L^{-1} (or mol dm^{-3}). However, the acidity is normally measured using the logarithmic pH scale (to base 10), by taking the log of the *numerical value* of $[H^+]$:

$$pH = -\log([H^+]) \qquad [5.19]$$

Example 5.9 compares the logarithmic pH scale with the direct concentration scale.

Example 5.9

Values of $[H^+]$ and pH for pure water and *example* values for a strong acid and a strong base:

	Strong acid (example)	Pure water	Strong base (example)
Hydrogen ion concentration: $[H^+]$/mol L^{-1}	0.1	1.0×10^{-7}	1.0×10^{-13}
pH value: $pH = -\log[H^+]$	$= -\log(0.1)$ $= 1.0$	$= -\log(1.0 \times 10^{-7})$ $= 7.0$	$= -\log(1.0 \times 10^{-13})$ $= 13.0$

The logarithmic pH scale can be considered to be a *linear scale* in 'acidity'; the pH-value increases for decreasing acidity. The numbers 1.0, 7.0, 13.0 are a more convenient *linear* representation of 'acidity' than the absolute scale of hydrogen ion concentrations for the same 'acidities': 0.1, 10^{-7}, 10^{-13}.

Q5.10

Calculate the pH-values equivalent to the following values for $[H^+]$:

(i) $[H^+] = 3.4 \times 10^{-9}$ mol L^{-1} (ii) $[H^+] = 3.4 \times 10^{-4}$ mol L^{-1}

In spectrophotometric measurements, the absorbance, A, of a solution is related to its percentage transmittance, $T\%$, according to the equation:

$$A = -\log\left(\frac{T\%}{100}\right) \qquad [5.20]$$

Example 5.10 demonstrates the rearrangement of logarithmic equations using the rules developed in 3.2 and 3.4.

Example 5.10

(i) Calculate the hydrogen ion concentration, $[H^+]$, for a solution with pH = 8.3.
(ii) Calculate the transmittance, $T\%$, for an absorbance, $A = 0.36$.

Substituting in equations [5.19] and [5.20]:

$$8.3 = -\log([H^+]) \qquad\qquad 0.36 = -\log\left(\frac{T\%}{100}\right)$$

Swap the equations around to put the unknown on the LHS using Rule 2 (see 3.2):

$$-\log([H^+]) = 8.3 \qquad\qquad -\log\left(\frac{T\%}{100}\right) = 0.36$$

Change the signs of every term on both sides of the equation using Rule 3:

$$\log([H^+]) = -8.3 \qquad\qquad \log\left(\frac{T\%}{100}\right) = -0.36$$

Take the inverse of 'log' using Rule 6 (3.4.2) or equation [5.7]:

$$[H^+] = 10^{-8.3} \qquad\qquad \frac{T\%}{100} = 10^{-0.36}$$

Using a calculator to evaluate $10^{-8.3}$ and $10^{-0.36}$ (also using Rule 5 for $T\%$):

$$[H^+] = 5.01 \times 10^{-9} \text{ mol L}^{-1} \qquad\qquad \frac{T\%}{100} = 0.4365$$
$$T\% = 100 \times 0.4365 \Rightarrow 43.7\%$$

Q5.11

Using [5.19], calculate the hydrogen ion concentrations equivalent to the following values for pH:

(i) pH $= 9.2$ (ii) pH $= 3.2$

Q5.12

Using [5.20], calculate the missing values in the following table of equivalent values (∞ = infinite absorbance):

Percentage transmittance, $T\%$	0		1		50	
Absorbance, A	∞	3		1		0

5.2   Exponential Growth and Decay

5.2.1   Introduction

A growth (or decay) is said to be *exponential* if time appears as an *exponent* or *power* in the equation of growth.

There are many examples of exponential growth or decay systems in all branches of science, and different disciplines have devised different ways of quantifying very similar processes, e.g. half-life in radioactivity, the elimination constant for drug concentration and the amplification of a photomultiplier. However, in this unit we see how many simple growth (or decay) systems can be represented, and analysed, by using one common exponential equation using the base e.

5.2.2   Exponential systems

There are many systems in science where the *future change* in the system depends on the *current* state of the system. For example:

- A population (e.g. bacterial cells in blood system) may increase by a constant *proportion* of the *current population*.
- In radioactive decay, the *rate of decrease* in radioactivity is proportional to the *current level* of radioactivity.

Q5.13

A poor (but consistent) gambler loses *exactly half* of his remaining money, M, every week. He starts with $M_0 = £640$ at the beginning ($n = 0$) of the first week, and is down to $M_1 = £320$ at the end of the first week ($n = 1$) and down to $M_2 = £160$ at the end of the second week ($n = 2$).

For *each value* of n in the table below calculate:

(i) how much money, M_n, is left by *halving the previous value*
(ii) the value of M_n, using the equation $M = 640 \times 0.5^n$.

End of week $n =$	(0)	1	2	3	4	5	6	7
(i) Money, M_n (£) $=$	640	320	160					
(ii) $M_n = 640 \times 0.5^n =$	640							

The answers to Q5.13 show that a *proportionate* change can be mathematically represented by an exponential equation. The *exponential* decay for Q5.13 is reproduced in Figure 5.1.

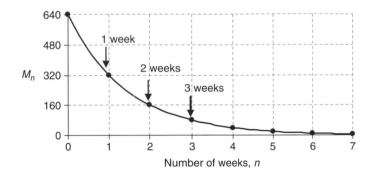

Figure 5.1 Exponential decay: $M_n = 640 \times 0.5^n$.

If a population increases by a **factor** g within a given **time period** T, then

$$\text{Growth factor, } g = \frac{\text{Population at end of period, } T}{\text{Population at start of period, } T}$$

Example 5.11

Calculate the growth factors, g, in the following situations:

(i) a population increases by 5 % every 10 years
(ii) a bacterial population decreases by 20 % every hour.

(i) Period, $T = 10$ years: $g = \dfrac{100\% + 5\%}{100\%} = \dfrac{105\%}{100\%} = 1.05$

A value of $g > 1$ shows a *growing* population.

(ii) Period, $T = 1$ hour: $g = \dfrac{100\% - 20\%}{100\%} = \dfrac{80\%}{100\%} = 0.8$

A value of $g < 1$ shows a *decaying* population.

Q5.14

Calculate the growth factors, g, in the following situations:

(i) A population doubles.
(ii) A population increases by 50 %.
(iii) A population falls by 10 %.
(iv) A population falls to 10 %.

When describing the growth (or decay) of a population we can write that:

Population at the start ($n = 0$) is given by N_0.
Population after n periods is given by N_n.

The sequence of populations below for each period of a growth shows that the population increases by a factor g every period, giving N_0, $N_0 \times g$, $N_0 \times g^2$, $N_0 \times g^3$, etc.

Period	Change in population		Population
0	N_0 (starting population)		$\rightarrow N_0$
1	$N_1 = N_0 \times g$	equivalent to:	$\rightarrow N_1 = N_0 \times g^1$
2	$N_2 = N_1 \times g$	substituting for N_1: $\quad N_2 = (N_0 \times g) \times g$	$\rightarrow N_2 = N_0 \times g^2$
3	$N_3 = N_2 \times g$	substituting for N_2: $\quad N_3 = (N_0 \times g^2) \times g$	$\rightarrow N_3 = N_0 \times g^3$
n		giving, after n steps: $N_n = (N_0 \times g^{n-1}) \times g \rightarrow$	$N_n = N_0 \times g^n$

Hence, after n periods, the population, N_n, is described by an exponential equation:

$$N_n = N_0 \times g^n \qquad [5.21]$$

Example 5.12

What are the values for the period T (in days) and the factor g in the example given in Q5.13?

The question states that after 1 week the money remaining has fallen to one-half.

Hence $g = 0.5$ for a period $T = 7$ days.

In general, the time, t, taken for n periods of duration T will be given by

$$t = n \times T \quad \text{and} \quad n = \frac{t}{T} \qquad [5.22]$$

Hence we can rewrite equation [5.21] directly in terms of t instead of n.
A population, N_t, which grows by a factor g in a time T is given by:

$$N_t = N_0 \times g^{t/T} \qquad [5.23]$$

Figure 5.2 shows the **exponential decay** of an initial population of $N_0 = 640$, which has a growth factor of $g = 0.5$ over a time period, $T = 7$ days (= 1 week). This is the same decay as in Figure 5.1, but the time, t, is now expressed in terms of *days* using equation [5.23].

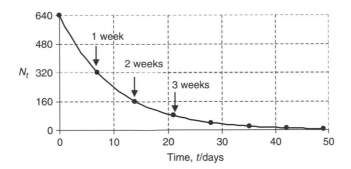

Figure 5.2   Exponential decay: $N_t = 640 \times 0.5^{t/T}$ with $T = 7$ days.

The equation, $N_t = N_0 \times g^{t/T}$, can be used to describe the growth (decay) of *any exponential system* over a time, t.

For a *particular* system it would be necessary to choose appropriate values for each of the variables, N_0, g and T.

Example 5.13

The equation for the growth of bacteria (see 5.2.4) can be written as $N_t = N_0 \times 2^{t/T_G}$, where T_G is the generation time of the bacteria.

In this case the bacteria numbers will double (growth factor, $g = 2$) within the period, T_G = generation time.

Q5.15

In a 'chain letter', one person sends a letter to six people who are *each* 'encouraged' to forward the letter to six new people within 2 weeks.

Assume that the chain is not broken and all six people actually forward the letter to *new* recipients.

(i) What is the 'period' of change in this example?
(ii) What is the value for g in this example?
(iii) Using an equation of the form, $N_t = N_0 \times g^{t/T}$, estimate how many weeks, months or years it would take before at least 20 million people have become involved in the chain.

Q5.16

A bacterial culture in the 'death phase' has an initial population of 2.0×10^6 cells per mL, which decays exponentially to one-tenth of its initial population in 1.2 hours.

(i) Calculate the population after 3.6 hours.
(ii) What is the value for g in this example?
(iii) Using an equation of the form, $N_t = N_0 \times g^{t/T}$, calculate the population after 2.2 hours.

5.2.3   Exponential growth equation $N_t = N_0 e^{kt}$

Growth and decay in different areas of science have produced different expressions of the basic equation $N_t = N_0 \times g^{t/T}$. However, the problem with this form is that different examples may need different values for the *base*, g.

It is more *convenient* to use a *standard equation* that always uses the same 'base'. For reasons that may not be obvious at first, the best choice for a common 'base' is Euler's number, e:

$$N_t = N_0 e^{kt} \qquad\qquad [5.24]$$

The base, e, is used extensively because of its simplicity when used in the *differentiation* (6.2.2) of growth and decay equations.

In the above equation, we replaced the growth factor, g, and time period, T, with a single exponential growth factor, k, whose value will depend on the value of g. Since the product, $k \times t$, is a pure number, k will have units of '1/time'.

Comparing equations [5.23] and [5.24], the two equations will be identical if:

$$e^{kt} = g^{t/T}$$

Taking natural logarithms of both sides:

$$\ln(e^{kt}) = \ln\left(g^{t/T}\right)$$

Using equation [5.8], the above equation becomes:

$$kt = \frac{t}{T} \times \ln(g)$$

Cancelling t from both sides, we get:

$$k = \frac{\ln(g)}{T} \text{ and then } T = \frac{\ln(g)}{k} \tag{5.25}$$

5.2.4 Specific applications of $N_t = N_0 e^{kt}$

We identify four examples of using specific time periods in growth and decay:

1. The exponential growth of a bacterial population is often measured by the **generation time, T_G**, which is the time that the population takes to *double* in number.
2. The decay of a radioactive isotope is measured by its **half-life, $T_{1/2}$**, which is the time that the radioactivity takes to fall to *half* of its initial value.
3. The decay of signals (particularly in electronics) is often described by a **time constant, τ (tau)**, which is the time taken to drop to a *fraction*, e^{-1}, ($\sim$36.8 %) of its initial value.
4. The decay of a bacterial population is often described by the **decimal reduction time, T_D**, which is the time taken to fall to *one-tenth* of its initial value.

Values for k can be calculated from g and T for the four applications using [5.25] and are given in Table 5.1.

Table 5.1. Specific expressions for the growth factor, k.

Time period	g	k	Equation	Applications
Generation time	2.0	$k = \dfrac{0.693}{T_G}$	$N_t = N_0 e^{0.693 \times t / T_G}$	Bacterial growth
Half-life	0.5	$k = -\dfrac{0.693}{T_{1/2}}$	$N_t = N_0 e^{-0.693 \times t / T_{1/2}}$	Radioactive decay
Decrease to $e^{-1} = 36.8\%$	0.368	$k = -\dfrac{1.000}{\tau}$	$N_t = N_0 e^{-t/\tau}$	Time constant
Decimal reduction time	0.1	$k = -\dfrac{2.303}{T_D}$	$N_t = N_0 e^{-2.303 \times t / T_D}$	Bacterial decay

These calculations show that, for an exponential *decay*, g will be less than 1.0, and k will be *negative*.

Example 5.14

A bacterial colony has an initial population of 3.6×10^3 cells and a generation time, $T_G = 25$ minutes, calculate the population after:

(i) 50 minutes
(ii) 80 minutes

(i) From the definition of 'generation time', the population will double ($g = 2$) after 25 minutes, and then it will double again in the next 25 minutes, so that after 50 minutes the population will be:

$$N_{50} = 3.6 \times 10^3 \times 2 \times 2 \Rightarrow 14.4 \times 10^3 \Rightarrow 1.44 \times 10^4 \text{ cells}$$

(ii) The time, $t = 80$ minutes, is not a simple multiple of the generation time, and so we cannot now use the quick method as in (i).
However, we can first calculate the value for k with equation [5.25]:

$$k = \ln(g)/T_G \Rightarrow \ln(2)/25 \Rightarrow 0.693/25 \Rightarrow 0.02773 (4 \text{ sf})$$
$$N_0 = 3.6 \times 10^3$$

Substituting into equation [5.24], the population, N_{80}, after 80 minutes is:

$$N_{80} = 3.6 \times 10^3 \times e^{0.02773 \times 80} \Rightarrow 3.31 \times 10^4 (3 \text{ sf})$$

Q5.17

The radioactive isotope radon-222 has a half-life of 3.8 days. If a sample has an initial radioactivity of 10.8 s^{-1}, calculate the radioactivity after:

(i) 11.4 days
(ii) 10.0 days

Example 5.15

Calculate the decimal reduction time T_D for a bacterial colony that decreases from 6.0×10^4 cells to 2.0×10^3 in 42 minutes.

The analysis is performed in the following text.

In Example 15.5, we use [5.24] and the expression for k in Table 5.1, to write directly:

$$N_t = N_0 \exp(-2.303 \times t/T_D)$$

where t is the time in minutes.
  Substituting the values for $N_0 = 6.0 \times 10^4$ and $N_{42} = 2.0 \times 10^3$ for $t = 42.0$:

$$2.0 \times 10^3 = 6.0 \times 10^4 \times \exp(-2.303 \times 42.0/T_D) \Rightarrow 6.0 \times 10^4 \times \exp(-96.73/T_D)$$

swapping the equation around (3.2.1) and dividing both sides by 6.0×10^4 (3.2.3) gives:

$$\exp(-96.73/T_D) = 2.0 \times 10^3/6.0 \times 10^4 \Rightarrow 0.0333$$

Taking the inverse of the exponential (3.4.2):

$$-96.73/T_D = \ln(0.0333) \Rightarrow -3.402$$

Dividing both sides by -96.73 (3.2.3) and taking the reciprocal of both sides (3.4.2):

$$T_D = -96.73/-3.402 = 28.4 \text{ minutes}$$

This decay is plotted in Figure 5.3.
  The calculation in Example 5.15 gives the decimal reduction time T_D as 28.4 minutes, and we can see in Figure 5.3 that, when $t = 28.4$ minutes, the number of cells has decreased to 6.0×10^4. This value is *one-tenth of the initial value*, which agrees with the definition of decimal reduction time.

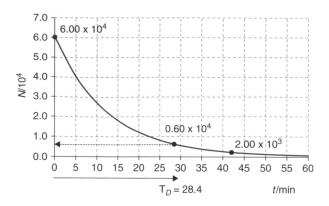

Figure 5.3 Exponential decay in Example 5.15.

Q5.18

Calculate the generation time T_G for a bacterial colony that increases from 2.0×10^4 cells to 1.7×10^6 in 180 minutes.

Example 5.16

A radioactive isotope decays with a half-life $T_{1/2} = 30$ days. Calculate:

(i) the proportion (fraction), F_{100}, of activity remaining after 100 days
(ii) the time taken for the activity to fall to 1 % of its initial value.

(i) Using [5.24] and the expression for k in Table 5.1, we can write directly:

$$N_t = N_0 \exp(-0.693 \times t / T_{1/2}) \Rightarrow N_0 \exp(-0.0231 \times t)$$

where t is the time in days.
The remaining *proportion* of activity is given by:

$$F_t = \frac{N_t}{N_0} \Rightarrow \frac{N_0 \exp(-0.0231 \times t)}{N_0} \Rightarrow \exp(-0.0231 \times t)$$

After 100 days:

$$F_{100} = \exp(-0.0231 \times 100) \Rightarrow 0.0993$$

(ii) An activity of 1 % of the initial value after a time, t, equals a proportion:

$$F_t = N_t / N_0 = 0.01$$

Which, substituting into the equation in (i), gives:

$$0.01 = \exp(-0.0231 \times t)$$

Swapping the equation from side to side:

$$\exp(-0.0231 \times t) = 0.01$$

Taking the inverse operation of the power of e:

$$-0.0231 \times t = \ln(0.01) \Rightarrow -4.605$$

Hence:

$$t = -4.605/-0.0231 = 199.4 \text{days}$$

Q5.19

The activity of an unknown radioactive isotope, X, is found to decay to one-tenth of its initial activity after a period of 26 hours.

Calculate the half-life of the isotope.

Q5.20

The time constant, τ, of a capacitor of capacitance C which discharges through a resistor of resistance R is given by:

$$\tau = CR$$

Calculate the time taken for a capacitor to discharge to 1% of its initial charge, given that $C = 0.01 \times 10^{-6}$ F and $R = 330 \times 10^{3} \Omega$.

5.2.5   General use of the equation $N_t = N_0 e^{kt}$

A common problem requires the calculation of the constant, k, when particular values of N_0, N_t and t are known. The solution is obtained by substituting the known values into [5.24] and then rearranging the equation to calculate the unknown, k.

Example 5.17

A bacterial population, initially with $N_0 = 5.2 \times 10^5$ cells per mL, is decaying exponentially according to the equation $N_t = N_0\, e^{kt}$, and it is found that after $t = 2.0$ hours, the population has become $N_{2.0} = 3.5 \times 10^4$ cells per mL.

(i) Calculate the value of k in the equation.
(ii) Hence calculate the expected population after 3.0 hours.

(i) Substituting the known values for $t = 2.0$ into [5.24]:

$$3.5 \times 10^4 = 5.2 \times 10^5 \times e^{k \times 2.0}$$

Swapping the equation around and dividing both sides by 5.2×10^5:

$$e^{k \times 2.0} = 3.5 \times 10^4/5.2 \times 10^5 \Rightarrow 0.06731$$

Using equation [5.7] to take the inverse operation:

$$k \times 2.0 = \ln(0.06731) \Rightarrow -2.699$$

which gives:

$$k = -1.349 \text{ h}^{-1}$$

(ii) Now we can substitute the values of N_0, k and $t = 3.0$ to get:

$$N_{3.0} = 5.2 \times 10^5 \times e^{-1.349 \times 3.0} \Rightarrow 9.09 \times 10^3 \text{cells per mL.}$$

Q5.21

The population of a bacterial colony is initially measured as 450 and, after 10 hours, has grown to $N_{10} = 620$.

(i) Assuming that the growth is exponential, find the values N_0 and k that will be required to describe this growth when using the equation:

$$N_t = N_0\, e^{kt}$$

Note that the units of k in this problem will be 'h^{-1}'.
(ii) Estimate the population after 12 hours.

Q5.22

The exponential growth of a population, N, can be written as an equation:

$$N_t = N_0\, e^{kt}$$

where N_t is the population at time t days, and N_0 is the initial population at time $t = 0$ days. If $N_0 = 3500$ and $k = 0.02\ \text{day}^{-1}$, calculate the population after:

(i) 25 days   (iii) 75 days
(ii) 50 days

Q5.23

A disease is spreading exponentially, such that the number of cases is increasing by 10 % every week. If there are 100 cases at the beginning of the first week (when $t = 0$), derive an equation of the form $N_t = N_0\ e^{kt}$ to give the number of cases, N_t, after a time, t, measured in weeks.

Calculate the values for N_0 and k appropriate to this problem.

Note that the units of k in this problem will be 'week^{-1}'.

An equation containing an exponential term can sometimes approach a *non-zero* value. An example is illustrated by Q5.24 in which the value, V_t, starts at zero and then approaches a constant value, V_0, with a time constant, τ. The rate of change decreases as it approaches the constant value.

Q5.24

The following equation represents a form of growth that reaches a maximum saturation level, V_0.

$$V_t = V_0 \times \left[1 - \exp\left(-\frac{t}{\tau}\right)\right]$$

(i) For $V_0 = 10.0$ and $\tau = 1.5$, plot the behaviour of V_t on a graph against values of time, $t = 0, 1, 2, 3$ and 4.
(ii) From the graph, *estimate* values for the ratio (V_t/V_0) when $t = \tau$ and when $t = 2\tau$.
(iii) Use the equation to *calculate* the values for the ratio (V_t/V_0) when $t = \tau$ and when $t = 2\tau$.

It is also common to find exponential change occurring with a variable other than time, t.

Q5.25

The penetration of light intensity, I_d, vertically downwards at a depth, d(m), into a lake is given by the equation:

$$I_d = 25.0 \, e^{-kd}$$

If the light intensity $I_5 = 5.6$ lumens (lm) at a depth $d = 5.0$ m, calculate the light intensity I_{10} at a depth $d = 10.0$ m.

5.2.6 Linearizing $N_t = N_0 \, e^{kt}$

There are many occasions in science when we want to find the 'best fit' of a theoretical equation to a set of experimental data.

The procedure for finding the best-fit for straight line data is straightforward using both graphical and linear regression methods – see 4.2. However, it is far more difficult to calculate a best-fit equation for an exponential curve.

We will develop a compromise procedure for the exponential equation [5.24]:

$$N_t = N_0 \, e^{kt}$$

where we manipulate the exponential equation so that the data can be plotted as a straight line using appropriate co-ordinate axes.

The **linearization procedure** (see also 4.3.1) is to take logarithms of both sides of the equation. It is possible to use either 'ln' or 'log':

$$\ln(N_t) = \ln(N_0 \times e^{kt}) \quad \text{or} \quad \log(N_t) = \log(N_0 \times e^{kt})$$

Using equation [5.10] for the logarithm of a product, each equation becomes:

$$\ln(N_t) = \ln(N_0) + \ln(e^{kt}) \quad \text{or} \quad \log(N_t) = \log(N_0) + \log(e^{kt})$$

For linearization using the *natural logarithm*, equation [5.8] gives $\ln(e^{kt}) = kt$, and we find:

$$\ln(N_t) = \ln(N_0) + k \times t \qquad [5.26]$$

which is of the same form as the straight line equation:

$$y = c + m \times x$$

If we plot $\ln(N_t)$ as the y variable and t as the x variable we will get a straight line with:

Slope, $m = k$

Intercept, $c = \ln(N_0)$

k can be derived *directly* from the slope : $k = m$ [5.27]

N_0 can be *calculated* from the intercept, c, using $N_0 = e^c$ [5.28]

For linearization using *logarithms to base 10*, equation [5.9] gives:

$$\log(e^{kt}) = kt \times \log(e) \Rightarrow 0.4343 \times k \times t$$

and we find that the linearization equation from above becomes:

$$\log(N_t) = \log(N_0) + 0.4343 \times k \times t \qquad [5.29]$$

Plotting $\log(N_t)$ against t will still give a straight line as for $\ln(N_t)$, but with different coefficients:

Slope, $m = 0.4343 \times k$

Intercept, $c = \log(N_0)$

which can be rearranged to give:

$$k = 2.303 \times m \quad \text{and} \quad N_0 = 10^c \qquad [5.30]$$

The specific period, T, of the growth/decay (e.g. generation time, T_G) can then be calculated from k by using the equation [5.25] or the equations in Table 5.1.

Example 5.18

Experimental measurements of the population, N_t, of the dying bacterial colony described in Example 5.15 are recorded. Then, taking logarithms, the values of $\log(N_t)$ are plotted against the time, t (minutes).

A linear regression calculation (e.g. using the INTERCEPT and SLOPE functions in Excel) gives a best-fit line with:

$$\text{Initial value (intercept)}, c = \log(N_0) \Rightarrow 4.78$$

$$\text{Slope}, m = -0.0352\text{min}^{-1}$$

Calculate the best estimate for the decimal reduction time, T_D, for this colony.

The analysis is performed in the following text.

The data in Example 5.18 is plotted (see Figure 5.4) using $\log(N_t)$ so that we use $k = 2.303 \times m$ from [5.30], giving:

$$k = 2.303 \times (-0.0352) = -0.0811$$

The value for T_D is calculated from k by rearranging the equation $k = -2.303/T_D$ from Table 5.1:

$$T_D = \frac{-2.303}{k} \Rightarrow \frac{-2.303}{-0.0811} \Rightarrow 28.4 \text{ minutes (as in Example 5.15)}$$

Note that, when plotting $\log(N_t)$ against the time, t, the decimal reduction time, T_D, is equal to the negative reciprocal of the slope, $T_D = -1/m = 1/(-0.0352) = 28.4$.

The 'linearized' decay in Figure 5.4 should be compared with the *same* data when plotted, in Figure 5.3, as N_t directly against t.

It can be seen from Figure 5.4 that after one 'decimal reduction time', when $t = 28.4$ minutes, the value of $\log(N_t)$ has fallen from 4.78 to 3.78, i.e. a fall of 1.0 on the logarithm

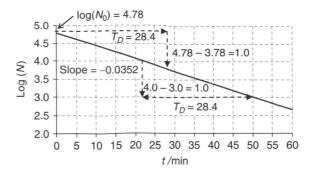

Figure 5.4 'Linearized' exponential decay in Example 5.18.

scale. It is also possible to see that this *rate of fall* is the same at any point on the graph, e.g. the value of $\log(N_t)$ also falls by 1.0 from 4.0 to 3.0 over the period from 22.2 to 50.6 minutes = 28.4 minutes.

Using a plot of the logarithmic (to base 10) decay, the 'decimal reduction time', T_D, can be recorded directly by noting **the time that the graph takes to fall by 1.0** on the $\log(N_t)$ scale.

Q5.26

In an experimental measurement, the activity, A_t, of radon-220 (thoron) is plotted as $\ln(A_t)$ against the time, t (seconds).

A regression analysis of the data gives:

$$\text{Slope} = -0.012 \text{ s}^{-1}$$

Calculate the best estimate for the half-life, $T_{1/2}$, for this isotope.

Q5.27

Most pharmokinetic processes are described by first-order kinetics. The drug concentration, C, in the body falls with time, t, according to the equation:

$$C = C_0 \, e^{-Kt}$$

where K is the elimination constant and C_0 the concentration at zero time.

In an experimental measurement, values of concentration, C, are measured as a function of time, t (hours), and a graph of $\ln(C)$ versus t is plotted.

A regression analysis of the data gives:

$$Slope = -0.61$$

$$Intercept = 4.1$$

(i) Calculate the value of the elimination constant, K.
(ii) Calculate the value of the initial concentration, C_0.

More complex equations sometimes involve the need to perform further changes to the plotted variables (4.3.3).

Example 5.19

In the Arrhenius equation, the rate constant, k, of a chemical reaction is a function of the activation energy, E (J mol^{-1}) and absolute temperature, $T(K)$:

$$k = Ae^{\left(-E/RT\right)}$$

where R is the gas constant ($= 8.31$ J mol^{-1} K^{-1}) and A is a constant.

In an experiment, the rate constant, k, is measured as a function of different temperatures, T. A graph of $\ln(k)$ is plotted against $1/T$ and a linear regression calculation gives a 'best-fit' slope, $m = -6450$.

Calculate the activation energy, E, for this reaction.

The first step in linearizing the Arrhenius equation is to take natural logarithms of both sides:

$$\ln(k) = \ln(A) - \frac{E}{RT}$$

If we now plot $\ln(k)$ as the y variable and write $(1/T)$ as the x variable:

$$\ln(k) = \ln(A) - \frac{E}{R}\left(\frac{1}{T}\right)$$

the slope of the resultant straight line will be:

$$m = -\frac{E}{R}$$

As the measured value of $m = -6450$, we can rearrange the equation to give:

$$E = 6450 \times R \Rightarrow 6450 \times 8.31 \Rightarrow 53.6 \times 10^3 \; J \; mol^{-1} \Rightarrow 53.6 \; kJ \; mol^{-1}$$

6
Rates of Change

Overview

 Website
- 'How to do it' video answers for all 'Q' questions.
- Excel files appropriate to selected 'Q' questions and Examples.

Nature is not static, and much interesting science is concerned with how systems change.

We start the first unit with familiar calculations of average speed, by recording change of position within a finite time interval. The calculation of speed is related to the slope of a line on a *distance* against *time* graph. We then develop the more general case for calculating speed at an instant, when the speed is represented by a continuously varying curve on a graph. This 'rate of change' mathematics is also applied to other systems.

Simple examples in Excel are used to demonstrate how continuously varying data can be *modelled* in a suitable spreadsheet format. The calculations for approximate models are relatively simple, but it is often possible to develop solutions to the required degree of accuracy.

The analytical mathematics (calculus) of differentiation is beyond the scope of this book. However, the second unit introduces the concept of 'differentiation' and gives a brief introduction to some ideas of calculus.

6.1 Rate of Change

6.1.1 Introduction

The most familiar 'rate of change' is probably *speed*. We express speed, v, as a rate of change of distance, z, with time, t, e.g. 'miles per hour' or 'metres per second'. However, 'rates of change' are found in all areas of science involving a variety of different variables.

We start with simple examples of average speeds within finite periods of time, and then progress to the more general case of a continuously varying $x-y$ relationship.

Essential Mathematics and Statistics for Science 2nd Edition   Graham Currell and Antony Dowman
Copyright © 2009 John Wiley & Sons, Ltd

6.1.2 Rate of change with time

We can express a 'rate of change' with a simple equation:

$$\text{Average speed, } v = \frac{\text{Change of distance}}{\text{Change of time}} = \frac{\Delta z}{\Delta t} \qquad [6.1]$$

Note that:

Change = Final value − Initial value

The Greek symbol Δ (capital delta) is often used to denote a 'difference' in value.

Example 6.1

A motorcyclist is initially at a distance of 5000 m (5 km) along a road from a petrol station.

(i) She drives *away* from the petrol station reaching a distance of 9000 m (9 km) after 200 s. Calculate her *average* speed in this period.
(ii) She continues to drive *away* from the petrol station, but at a faster speed, reaching a distance of 12 000 m (12 km) after a further 100 s. Calculate her new *average* speed in this period.
(iii) She then turns round and drives *towards* the petrol station, reaching it after a further 400 s. Calculate her average speed in this last period.

The analysis is performed in the following text.

Figure 6.1 uses a z–t diagram to represent the distance along the road, z, of the motorcyclist in Example 6.1. The straight lines, which connect the points, assume that the motorcyclist travels at a constant speed in each section of the journey.

(i) The first section of the journey is illustrated by triangle A in Figure 6.1, where:
Change in distance, $\Delta z = 9000 - 5000 = 4000$ m
Change in time, $\Delta t = 200$ s
The average speed in the first section is calculated using [6.1]:

$$v = \frac{\Delta z}{\Delta t} = \frac{4000}{200} = 20 \text{ m s}^{-1}$$

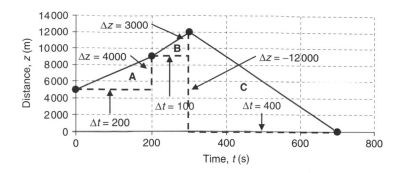

Figure 6.1 Distance–time graph for Example 6.1.

(ii) The second section of the journey is illustrated by triangle B in Figure 6.1, where the average speed is calculated using [6.1]:

$$v = \frac{\Delta z}{\Delta t} = \frac{12\,000 - 9000}{300 - 200} = \frac{3000}{100} = 30 \text{ m s}^{-1}$$

(iii) In the third section of the journey, illustrated by the triangle C in Figure 6.1, the change in distance is *negative*, which gives a *negative* average speed:

$$v = \frac{0 - 12\,000}{400} = \frac{-12\,000}{400} = -30 \text{ m s}^{-1}$$

The speed is *negative* in (iii) because the motorcyclist is travelling in a direction *opposite* to the direction in which the distance is being measured.

Example 6.2

Calculate the *slopes* of the straight line graph for each of the three triangular sections in the graph in Figure 6.1

The slope, m, of the straight lines are calculated using [4.8], giving:

Triangle A: $m = \dfrac{9000 - 5000}{200 - 0} = \dfrac{4000}{200} = 20 \text{ m s}^{-1}$

Triangle B: $m = \dfrac{12\,000 - 9000}{300 - 200} = \dfrac{3000}{100} = 30 \text{ m s}^{-1}$

Triangle C: $m = \dfrac{0 - 12\,000}{400} = \dfrac{-12\,000}{400} = -30 \text{ m s}^{-1}$

Comparing the results in Example 6.1 with those in Example 6.2, it can be seen that:

Rate of change of z with t equals the **slope of the graph** of z against t.

Q6.1

Calculate the missing values in the table below:

Initial distance (m)	Final distance (m)	Initial time (s)	Final time (s)	Speed (m s^{-1})
2000	4000	100	150	40
2000		100	200	30
2000	6000		400	40
2000		100	200	−40
5000	3000	0	200	

6.1.3 Modelling a continuous curve

In Example 6.1, the data is given at *discrete intervals*. We calculate an average speed between the points by assuming a constant (straight line) speed between those points.

For a curve with a constantly changing slope we can still measure changes over finite intervals to produce an *approximate* model for the variation in slope.

For example, in Figure 6.2 the slope at the point Q can be found by taking the differences, Δy and Δx, between two points, P and R, on either side of Q.

In Figure 6.2, the slope at Q, m_Q, is given approximately by:

$$m_Q \approx \frac{\Delta y}{\Delta x} \qquad [6.2]$$

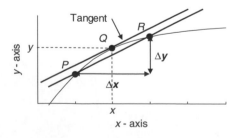

Figure 6.2 Approximation for rate of change.

If the curve is described by the co-ordinates of a series of points, then the slope of the curve can be calculated approximately by repeating the calculation in [6.2] between successive pairs of data points. This form of calculation can be *modelled* conveniently using Excel.

Example 6.3 illustrates the use of repeated calculations to *model* the change in surface gradient on a path that crosses a hill peak.

Example 6.3

Figure 6.3 is a map of height contours around a hill peak. Adjacent contours differ in height by 20 m, with the outer contour at a height $h = 60$ m. A straight path, given as the x-axis (in km), passes directly over the hill peak.

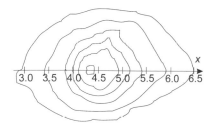

Figure 6.3   Height contours around a hill peak.

Calculate (approximately) the slope of the ground at the points where each contour (except the outer) crosses the x-axis.

The analysis is performed in the following text.

The positions where each contour crosses the x-axis in Example 6.3 are given in Table 6.1, as an extract from an Excel worksheet:

Table 6.1.   Heights along x-axis in Example 6.3.

A	B	C	D	E	F	G	H	I	J	K	L	M
4 Height, h (m)	60	80	100	120	140	160	160	140	120	100	80	60
5 Distance, x (km)	2.9	3.4	3.8	4.0	4.1	4.3	4.4	4.9	5.2	5.4	5.9	6.5
6 Slope, m (m km^{-1})		44	67	133	133	67	−33	−50	−80	−57	−36	

The calculation of slope at the position where $x = 3.4$ km is achieved by using the data points on either side, $x = 2.9$ and $x = 3.8$, and entering appropriate differences into [6.2] (the results are formatted to integer values):

$$m_{x=3.4} \approx \frac{\Delta h}{\Delta x} = \frac{(100 - 60)}{(3.8 - 2.9)} = 44.4 \text{ m km}^{-1}$$

The entry in the Excel worksheet for cell C6 is '=(D4-B4)/(D5-B5)'.

The slopes at each subsequent point are calculated by repeating the above calculation, moving from point to point. This form of repetitive calculation is very easily performed in Excel.

Notice that the slope is positive leading up to the peak and then negative after the peak.

Example 6.4

The following data shows the results of a potentiometric titration, where the cell potential is recorded as more titrate is added.

Calculate the 'end-point' of the titration, where the *rate of change of cell potential is greatest*.

Volume added	V (cm³)	8.0	8.5	9.0	9.2	9.4	9.6	9.8	10.0	10.2	10.4	10.6	10.8	11.0	11.5	12.0
Cell potential	E (mV)	154	161	173	178	188	197	211	228	237	248	255	260	265	269	273
Slope	$\dfrac{\Delta E}{\Delta V}$		19.0	24.3	37.5	47.5	57.5	77.5	65.0	50.0	45.0	30.0	25.0	12.9	8.0	

The rate of change (or slope) of the cell potential with volume is calculated using an equation derived from [6.2]. For example, the slope when volume added, $V = 8.5$ mL, is calculated:

$$\text{Slope (at } V = 8.5) \approx \frac{\Delta E}{\Delta V} = \frac{173 - 154}{9.0 - 8.0} = 19.0 \text{ mV cm}^{-3}$$

The graphs in Figure 6.4 show the cell potential and its slope plotted against the volume added.

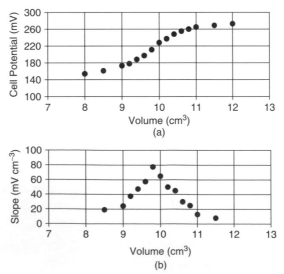

Figure 6.4 (a) Cell potential versus volume. (b) Slope versus volume.

The volume, $V = 9.8$, where the slope is greatest, can be seen most easily from the graph of the **slope** of the curve, at the point where this graph reaches a maximum.

Q6.2

The data in the table below gives the speed of a car as a function of time, as it accelerates to overtake a slower vehicle.

Take **Acceleration = Rate of change of speed with time**, and use [6.2] to estimate the acceleration between the times 2 and 28 seconds inclusive, and plot your results on a suitable graph.

Time (s)	0	2	4	6	8	10	12	14	16	18	20	22	24	26	28	30
Speed (m s^{-1})	20	20	20	21	23	27	31	34	35	35	35	34	33	32	32	32
Acceleration (m s^{-2})																

6.1.4 Slope of a graph at a point

We can aim to measure the slope of the graph at *a particular point* by making the measurement interval *as small as possible*. By making the measurement interval 'infinitely' small, we can then obtain a value for the slope of the curve at that point.

Figure 6.2 shows a continuous x–y curve with three points on the curve, P, Q and R. The points P and R are used to estimate the slope of the curve around Q, and are separated by small differences Δx and Δy in the x- and y-directions respectively.

The **average** slope of the curve between P and R is given *approximately* by equation [6.2].

In Figure 6.2, a straight line drawn between the points P and R crosses the curve at the two points P and R. If we now bring P and R closer towards Q, we can imagine reaching a situation when the straight line *only just touches* the curve at the point Q.

The straight line that **just touches** a curve at a point Q, but *does not cross the curve*, is called the **tangent to the curve at the point Q**.

When we bring P and R closer towards Q, and make Δx approach zero, we rewrite the ratio, $\Delta y/\Delta x$, as, and the slope of the curve at Q is given by:

$$\text{Limit}_{\Delta x \to 0} \frac{\Delta y}{\Delta x} \Rightarrow \frac{dy}{dx} \qquad [6.3]$$

dy/dx (pronounced 'dee' y by 'dee' x) is called the **differential coefficient**, the **derivative**, or the **slope at the point**.

The value of the differential coefficient (or slope), dy/dx, will vary from point to point along a curve.

Q6.3

The graph in Figure 6.5 shows the population, N_t, of a bacterial colony as a function of time, t.

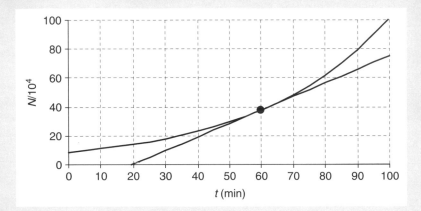

Figure 6.5

The population at time $t = 60$ min is 37.85×10^4, and the straight line is the tangent to the graph at that point.

(i) Using the drawn tangent, estimate the rate of growth of the colony at the time $t = 60$ min.
(ii) Draw a tangent to the curve at time $t = 90$ min, and hence estimate the rate of growth of the colony at the time $t = 90$ min.

6.2 Differentiation

6.2.1 Introduction

The concept of the differential coefficient was introduced in 6.1.4 as being the slope of a curve at a single point. In this unit we give some examples of equations that describe the slope of a function at any given point. These equations were derived using the mathematical processes of *calculus*. However, it is beyond the scope of this book to develop this advanced form of mathematics, but we include this section as a brief introduction to its power and uses.

We also see in 6.2.2 that an important property of Euler's number, e, is that the slope (or differential coefficient) of the curve $y = e^x$ is also equal to e^x. This is a unique property that only applies to the number e = 2.718 28 ..., and is the reason that 'e' is used so often in the mathematics of growth and decay.

6.2.2 Differentiation

Differentiation is the mathematical process (in calculus) for calculating the differential coefficient, dy/dx, *directly from the equation* that relates y to x. We introduce this concept using Example 6.5

Example 6.5

Plot the curve $y = x^3$ between $x = 0.0$ and 2.5.

(i) Draw tangents to the curve at points $x = 1.0$ and 2.0.
(ii) Calculate the slopes of the tangents in (i) from the graph.
(iii) Hence, show that the results in (ii) are consistent with the fact that the slope, $m(= dy/dx)$, of the curve $y = x^3$ is given by:

$$\frac{dy}{dx} = 3x^2$$

The analysis is performed in the following text.

Plotting the curve $y = x^3$ in Figure 6.6, we calculate the slope, $m_{x=1}$, when $x = 1$, and $m_{x=2}$, when $x = 2$, by drawing tangents at these points and measuring the slopes of these tangents. The slope of each tangent is calculated by using equation [4.8] applied to the points at both ends of each tangent.

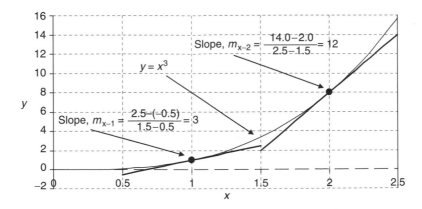

Figure 6.6 Slopes of $y = x^3$.

By construction, we find that when $x = 1$, $m_{x=1} = 3$, and when $x = 2$, $m_{x=2} = 12$. Using the equation for the slope (or differential coefficient), $dy/dx = 3x^2$:

$$\text{for } x = 1, \text{ slope } m_{x=1} = \left(\frac{dy}{dx}\right)_{x=1} = 3 \times 1^2 \Rightarrow 3$$

$$\text{for } x = 2, \text{ slope } m_{x=2} = \left(\frac{dy}{dx}\right)_{x=2} = 3 \times 2^2 \Rightarrow 12$$

We can see that our graphical construction is consistent with the equation for the differential coefficient in this particular problem.

In general, there are a number of rules in calculus for deriving the differential equation *from* the original equation. In this book we only give some simple examples – see Table 6.2.

Table 6.2.   Differentials of common equations (c, A and B are constants)

Equation	Differential coefficient	Comment
$y = c$	$\dfrac{dy}{dx} = 0$	The rate of change of a constant is zero
$y = Ax^B$	$\dfrac{dy}{dx} = ABx^{B-1}$	Differentiation of a power
$y = Ae^{Bx}$	$\dfrac{dy}{dx} = ABe^{Bx}$	Differentiation of an exponential

Example 6.5 is a 'power' equation, $y = Ax^B$, where $A = 1$ and $B = 3$. From Table 6.2 we can see that the differential coefficient of this equation would become (substituting for A and B):

$$\frac{dy}{dx} = ABx^{B-1} \Rightarrow 1 \times 3 \times x^{3-1} \Rightarrow 3x^2$$

If, in Table 6.2, we consider the simple exponential equation $y = e^x$, with $A = 1$ and $B = 1$, then its differential coefficient will be:

$$\frac{dy}{dx} = 1 \times 1 \times e^{1x} \Rightarrow e^x$$

Hence the differential coefficient of e^x is also e^x. This is the special property of 'e' which makes it such an important number in growth and decay equations.

Example 6.6 now illustrates the differentiation of an exponential equation to calculate the rate of growth of a bacterial colony.

Example 6.6

The number, N_t, in a bacterial colony with a generation time of 28 minutes can be described (5.2.4) using the following equation, where t is given in minutes:

$$N_t = N_0 e^{0.0247 \times t}$$

If the initial colony size is $N_0 = 8.6 \times 10^4$ (when $t = 0$), calculate the rate of growth (in cells per minute) when $t = 60$ minutes.

If we compare the equation with the general exponential equation $y = Ae^{Bx}$, we find the equivalences:

$$y \rightarrow N_t, A \rightarrow N_0, B \rightarrow 0.0247, x \rightarrow t$$

We see from Table 6.2 that the differential coefficient of the general exponential equation $y = Ae^{Bx}$ is then given by:

$$\frac{dy}{dx} = ABe^{Bx}$$

Substituting the values for y, x, A and B appropriate to this problem:

$$\frac{dN_t}{dt} = 8.6 \times 10^4 \times 0.0247 \times e^{0.0247 \times t}$$

Then substituting for $t = 60$ gives:

$$\left(\frac{dN_t}{dt}\right)_{t=60} = 9350 \text{ min}^{-1}$$

i.e. when $t = 60$, the rate of growth is 9350 cells per minute.

The rate of change with time is sometimes written with a 'dot' over the variable. For example, the cardiac output in terms of volume, V, of millilitres of blood per minute can be written as:

$$\dot{V} = \frac{dV}{dt} \text{mL min}^{-1}$$

Note that, when writing out the differential coefficient of an expression for y, it is often convenient to write the expression in a bracket after d/dx. For example, if $y = Ae^{Bx}$, then we would write:

$$\frac{dy}{dx} = \frac{d}{dx}(Ae^{Bx})$$

Example 6.7

The distance, z, travelled by a falling stone as a function of time, t (ignoring air resistance), is given by:

$$z = 4.9 \times t^2$$

Given that velocity = rate of change of distance with time, calculate the velocity of the stone after 2 seconds.

We know that:

$$\text{Velocity, } v = \text{Rate of change of distance, } \frac{dz}{dt} = \frac{d}{dt}(4.9 \times t^2)$$

Comparing with the standard equation from Table 6.2:

$$\frac{d}{dx}(Ax^B) = A \times B \times x^{B-1} = ABx^{B-1}$$

we can deduce, by putting $A = 4.9$, $B = 2$ and $x = t$, that:

$$v = \frac{d}{dt}(4.9 \times t^2) = 4.9 \times 2 \times t^{2-1} = 9.8 \times t$$

When $t = 2.0$, we can calculate that:

$$v_{t=2} = 9.8 \times 2.0 = 19.6 \text{ m s}^{-1}$$

6.2.3 Rate equations

There are many examples in science where the rate at which a quantity grows or decays is proportional to the quantity at that time, as illustrated in Example 6.8.

Example 6.8

A chemical A breaks down into other compounds B and C:

$$A \rightarrow B + C$$

In a first-order rate equation, the rate at which the concentration of A decreases with time, t, is proportional to the concentration of A at time t (usually written as $[A]_t$). Hence we can write:

$$\frac{d[A]_t}{dt} = -k[A]_t$$

where k is a constant and the negative sign shows a decay.

How will the decay of $[A]_t$ appear on a graph as a function of t?

The analysis is performed in the following text.

To answer Example 6.8, we need to find an expression for $[A]_t$ that makes the differential equation true (i.e. a solution to the equation).

With experience, we can anticipate that substituting $[A]_t = [A]_0 e^{-kt}$ will make the differential equation in the question balance. Using Table 6.2:

$$\frac{d[A]_t}{dt} = [A]_0 \times (-k) \times e^{-kt} \Rightarrow -k[A]_t$$

Hence the exponential decay $[A]_t = [A]_0 e^{-kt}$ is a true solution to the equation, and will describe the decay of chemical, A.

Q6.4

In microbiology, the rate of growth of a population of N_t organisms at a time t is given by:

$$\frac{dN_t}{dt} = kN_t$$

where k is a constant.

Derive an equation that describes the population N_t as a function of time, t.

Q6.5

The decay of a bacterial population is described by the equation:

$$N_t = 5.6 \times 10^8 \times \exp(-0.4 \times t)$$

where N_t is the population at a time t(s).

Calculate the number of bacteria dying per second at the following times (the differential coefficient will be negative as the numbers of bacteria are decreasing):

(i) $t = 0$ s   (iii) $t = 10$ s

(ii) $t = 5$ s

6.2.4   Estimating changes

If the value of the differential coefficient can be calculated, it is then possible to estimate the effect of *small* changes, Δx and Δy, by using the approximation:

$$\frac{\Delta y}{\Delta x} \approx \frac{dy}{dx} \qquad\qquad [6.4]$$

which can be rearranged to give:

$$\Delta y \approx \frac{dy}{dx} \times \Delta x \qquad\qquad [6.5]$$

Example 6.9

Using the results from Example 6.6, the rate of growth of a particular bacterial colony when $t = 60$ min is:

$$\left(\frac{dN_t}{dt}\right)_{t=60} = 9350 \text{ min}^{-1}$$

Estimate the increase in colony size, ΔN, between $t = 58$ min and $t = 62$ min.

The time interval, $\Delta t = 62 - 58 = 4$ min. Using [6.5] for this problem, we can write:

$$\Delta N \approx \left(\frac{dN_t}{dt}\right)_{t=60} \times \Delta t$$

giving:

$$\Delta N \approx 9350 \times 4 = 37\,400 \text{ cells}$$

Q6.6

The volumes, V (m^3), of adult animals of a particular species are related approximately to their heights, h (m), by the equation $V = 0.04 \times h^3$.

(i) Derive an expression for the differential coefficient, dV/dh.
(ii) Hence estimate the increase in volume (ΔV) of an animal of height $h = 1.6$ m, if it then grows by ($\Delta h =$) 0.02 m.

7

Statistics for Science

Overview

- 'How to do it' video answers for all 'Q' questions.
- Excel tutorial: statistical calculations.
- Excel and Minitab files appropriate to selected 'Q' questions and Examples.

All scientific experiments suffer from experimental uncertainty due to either subject variability and/or measurement variability (1.2). This variability can occur with the actual subjects being measured, e.g. different colonies of bacteria will grow at different rates even though the conditions appear to be the same. The variability can also appear in the actual measurement process itself, e.g. repeated measurements on the same fragment of glass may yield slightly different values for its refractive index, because of the limited accuracy of the measurement process.

There are three main ways in which statistical analysis is particularly useful in science:

- Presenting data in visual formats which aid understanding of the underlying patterns.
- Calculating 'best estimates' for the 'true' values of the parameters being measured.
- 'Testing' for the 'truth' of possible scientific hypotheses through an evaluation of the probabilities of error.

This chapter starts with a common situation in data analysis, i.e. repeated (replicate) measurements made to obtain a *best estimate* of an unknown value. The first step is to be able to display the data using a *box and whisker plot*. The next is to be able to calculate a quantitative result, the *confidence interval*, which incorporates an expression of the underlying uncertainty in the data.

The following units then develop the elements of statistics that are particularly useful for data analysis in science, introducing methods that are used to describe and quantify the variability that occurs in a simple set of experimental data. The section highlights the fact that a set of experimental results only represents a *sample* of all the possible values that could be obtained if the experiment could be repeated over and over again. We then use statistics to *infer*, from a data sample, *best estimates* for the characteristics of the actual system being measured.

With more extensive data sets it is often more convenient to record the number (or *frequency*) of data values falling within given *data ranges*. The characteristics of the data are then described by the characteristics of the overall *frequency* distribution or probability distribution.

Essential Mathematics and Statistics for Science 2nd Edition   Graham Currell and Antony Dowman
Copyright © 2009 John Wiley & Sons, Ltd

Finally, a section on factorials, permutations and combinations gives further basic statistics that underpin the later development of the binomial theorem and non-parametric tests.

Additional material on the Website includes the Bayesian approach to probability, which uses prior and posterior odds and likelihood ratios. Bayesian statistics are being used more extensively in managing probabilities in complex 'real-life' situations.

7.1 Analysing Replicate Data

7.1.1 Introduction

Experimental measurements are often made to discover the 'true' value of some parameter, e.g. the pH of a solution. However, all experimental measurements are subject to uncertainty, which means that the final value will only be a 'best estimate' for the unknown 'true' value. It is therefore common to make repeated (replicate) measurements to counteract the effects of random experimental uncertainties.

In this context, statistics can be used to perform two key functions:

1. Provide a description of the experimental 'raw' data that has been recorded, often presenting this in the form of a visual 'picture' of the distribution of the values.
2. Calculate, using certain statistical assumptions, a 'best estimate' of the 'true' value being measured.

We will use the experimental data in Example 7.1 to introduce the use of a *box and whisker plot* to describe the raw data and then we will use a case study to investigate the calculation of a *95 % confidence interval* to give the best estimate of the 'true' value being measured.

7.1.2 Ranked data – box and whisker plots

Example 7.1

The following data set has nine replicate experimental measurements of an unknown 'true' value, μ.

| Data | 2.3 | 11.3 | 3.8 | 4.5 | 4.2 | 8.1 | 6.3 | 3.7 | 3.3 |

Produce a visual representation of the values in this data set.

The worked answer is given the following text.

Ranked data has the data values *sorted* into ascending (or descending) order and assigned a *rank position*.

The nine data values in Example 7.1 can be sorted into ascending order and assigned rank positions from 1 to 9.

Data:	2.3	3.3	3.7	3.8	4.2	4.5	6.3	8.1	11.3
Rank:	1	2	3	4	5	6	7	8	9

Ranking is used extensively in *non-parametric* statistics (Chapter 12), where only the 'order' of the data values is important and not the actual (parametric) values.

The *location* of a set of, n, data values is described by the following:

- **Median** is the *middle* value in a set of ranked data values and gives the *location* of the data. The median is the value with the rank $\mathbf{0.50} \times (\mathbf{n + 1})$.
- **Lower quartile**, Q_1, is the value *one*-quarter of the way from the lowest to the highest value. The lower quartile is the value with the rank $\mathbf{0.25} \times (\mathbf{n + 1})$.
- **Upper quartile**, Q_3, is the value *three*-quarters of the way from the lowest to the highest value. The upper quartile is the value with the rank $\mathbf{0.75} \times (\mathbf{n + 1})$.

In the above data, the number of data values, $n = 9$. Thus:

- Median value has the rank $= 0.5 \times (9 + 1) = 5$
  Median value with rank $5 = 4.2$
- Lower quartile value has the rank $= 0.25 \times (9 + 1) = 2.5$
  Lower quartile value with rank 2.5 is halfway between 3.3 and $3.7 = 3.5$
- Upper quartile value has the rank $= 0.75 \times (9 + 1) = 7.5$
  Upper quartile value with rank 7.5 is halfway between 6.3 and $8.1 = 7.2$

The *spread* of non-parametric data is described by the following:

- **Interquartile range**, *IQR*, is the difference in value between the upper quartile and lower quartile:

$$IQR = Q_3 - Q_1$$

- **Total range** is the difference in value between the *lowest value* and the *highest value*.

Q7.1

For each of the sets below, calculate:

	Median	Lower quartile	Upper quartile	IQR
(i) 5.0, 8.2, 7.9, 6.6, 7.6, 5.7, 3.2, 7.5, 5.9				
(ii) 11, 8, 13, 9, 21, 24, 12, 22, 29, 43				
(iii) 45, 67, 23, 78, 67, 56, 98, 23, 49				

A box and whisker plot (often just called a boxplot) is very useful way of visualizing raw experimental data.

Figure 7.1 shows the box and whisker plot for the data in Example 7.1, drawn against the data value axis.

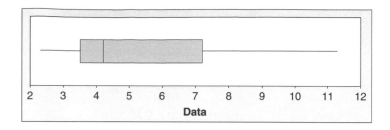

Figure 7.1   Box and whisker plots of data in Example 7.1 (using Minitab).

The *'middle' line* drawn inside the box shows the position of the *median* value.
The *ends of the 'box'* give the positions of the *upper and lower quartiles*.
The *ends of the 'whiskers'* give the *maximum and minimum* values in the data.

The fact that the median is not at the centre of its 'box' shows that the data is not symmetrical.

Q7.2

Represent all of the number sets in Q7.1 as box and whisker plots.

It is not possible to use standard Excel to draw box and whisker plots. However, statistics packages such as Minitab can produce these diagrams easily.

7.1.3   Confidence interval for an unknown value

If we are trying to measure some experimental property with a true value of μ, then we are more likely to record a value, x, close to μ, and less likely to record a value a long way away from μ. The graph in Figure 7.2 shows the likelihood distribution for getting an experimental result, x, when measuring a true blood–alcohol level of $\mu = 80$ mg of alcohol in each 100 mL of blood, i.e. 80 mg per 100 mL.

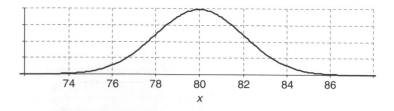

Figure 7.2   Distribution of possible experimental values.

In this example, we have assumed that most measurements fall between 78 and 82 and nearly all between 76 and 84, with only a few outside this range.

The mathematical name for this 'bell-shaped' distribution is the **normal distribution** (see 8.1.3). Most *random* experimental errors can be considered to have a frequency spread of results that follow a normal distribution.

Example 7.2

Twenty analysts each make five replicate measurements on a blood sample which has a true value of 80 mg per 100 mL. The individual recorded values are distributed at random with probabilities given by the curve in Figure 7.2.

How can their results be presented and interpreted?

The worked answer is given in the following text.

For Example 7.2, a random selection of values by computer has been used to simulate the realistic spread of 20 possible data sets, each with five experimental data values. The results are described by the 20 box and whisker plots in Figure 7.3.

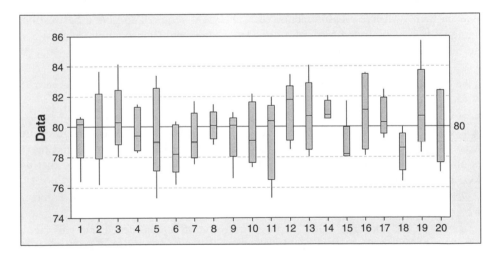

Figure 7.3 Box and whisker plots for 20 data sets (using Minitab).

It is important to realize that, in a real experimental measurement, none of the 20 analysts would know either the true value ($\mu = 80$) being measured or the uncertainty in the measurements. Individually, they must rely on just their five measurements to estimate both the true value and measurement uncertainty.

For example:

- Set 14 has data values that happen to be grouped closely together and all on the high side of the true value, and the analyst would tend to believe that his or her results are quite precise and close to a value of about 81.

- Set 5 has a wide spread of values and the analyst would not claim the same precision as analyst 14.
- Set 15 shows results that are skewed to low values (the median value is close to the lower end of the range) and set 1 is skewed to high values, whereas set 13 is fairly symmetrical.
- Set 18 only just includes the true value of 80.

The data for these six sets are given in Table 7.1.

Table 7.1. Data for analyst sets 1, 5, 13, 14, 15 and 18.

Set	Data					Mean	95 % confidence interval	
							Minimum	Maximum
1	80.2	79.6	80.7	80.4	76.4	79.5	77.3	81.6
5	81.7	83.4	78.9	75.3	79.0	79.7	75.8	83.5
13	78.0	79.0	81.7	80.7	84.1	80.7	77.7	83.7
14	80.8	82.1	80.5	80.6	81.3	81.1	80.2	81.9
15	78.1	81.7	78.3	78.2	78.0	78.9	76.9	80.8
18	78.6	79.1	76.4	80.0	77.9	78.4	76.7	80.1

Based on their own five measurements, each analyst is required to give a best estimate of the unknown true value in the form of a confidence statement. For example, analyst 5 would present his or her results as follows:

> On the basis of my five data values, I am 95 % confident that the true value, μ, lies between 75.8 and 83.5.

This range of values, called the 95 % confidence interval, is a symmetrical range centred on the mean (average) value of the set of sample data – see 8.2.4. The minimum and maximum values of the confidence interval in Table 7.1 have been calculated using theory that will be developed in Chapter 8.

The 95 % confidence intervals are calculated for each data set and recorded in Figure 7.4 as 'error bars' on either side of the sample mean values.

The statement that the 'true value, μ, lies within the 95 % confidence interval' has a 5 % chance of being *wrong*. Hence we could expect that 5 % of claims (i.e. 1 in 20) will indeed prove to be wrong. The simulated data agrees with this probability, in that it can be seen that just one confidence interval (from set 14) does not include the true value ($\mu = 80$).

Example 7.3

Which of the 20 analysts, whose results are given in Figure 7.4, would answer Yes to the following question:

> On the basis of your results, are you 95 % confident that the true blood–alcohol level is *not* 82 mg per 100 mL?"

We can see in Figure 7.4 that the confidence intervals for sets 1, 4, 6, 7, 8, 9, 14, 15 and 18 do *not* cross the 82 mg per 100 mL grid line.

As the '82' value is *outside* their separate confidence intervals for the true blood–alcohol level, each of *these* analysts would answer *yes* to the above question.

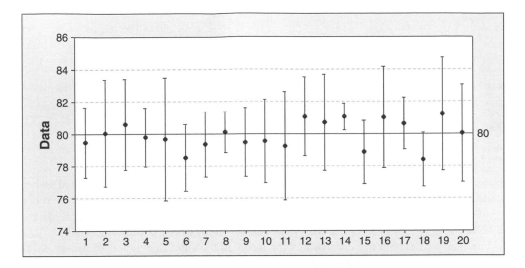

Figure 7.4   The 95 % confidence intervals for 20 data sets (using Minitab).

Q7.3

Case study data has been presented using both box and whisker plots in Figure 7.3 and confidence intervals in Figure 7.4.

Explain the fundamental difference between the information described by the two forms of presentation.

The difference between the two types of presentation can be illustrated by the effect of increasing the sample size of data values. The presentations in Figure 7.5 show a sample size $n = 5$ (set 1 from the data), sample size $n = 10$ (set 1 plus set 2), sample size $n = 20$ (sets 1 to 4), sample size $n = 40$ (sets 1 to 8) and sample size $n = 80$ (sets 1 to 16).

The interquartile ranges of the box and whisker plots (Figure 7.5a) approach a constant range which is representative of the underlying uncertainty in the data being measured, and the ends of the whiskers extend to the most extreme data values in the set.

However, the confidence interval (Figure 7.5b) is a measure of confidence in locating the true value being measured, and as the sample size increases, the increased information means that it is possible to be more precise about the true value – the confidence interval becomes narrower. This effect due to the central limit theorem is explained in 8.2.3.

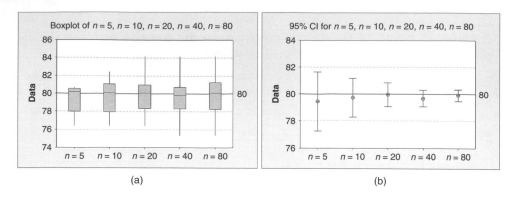

Figure 7.5   Effect of increasing sample size on (a) boxplots and (b) confidence intervals (using Minitab).

7.2   Describing and Estimating

7.2.1   Introduction

We saw in 7.1 that a set of replicate experimental measurements could be considered as being a statistical *sample* of the much larger *population* of all the replicate measurements that could be made. For example, the calculation of confidence intervals based on the sample data can then be used as an *estimate* for the true value for the mean of the population.

In this unit we develop the calculations for mean and standard deviation of a data set, and we start by addressing an important question:

> Are the values in the set to be considered as a *self-contained set* of numbers (a *population*), or are they *representative* (a *sample*) of a bigger set of other possible measurements (*source population*)?

The statistics that must be used for *samples* and *populations* are subtly different.

The results of an experiment are normally just a sample of the many different results that could be obtained if the experiment were to be repeated. Hence we normally apply *sample statistics* to experimental data and not population statistics.

7.2.2   Populations and samples

We start by considering the data set given in Example 7.4.

Example 7.4

Describe the *characteristics* of the data set A:

A	31	36	44	39	40

The worked answer is given in the following text.

The way in which we treat the five numbers in Example 7.4 will depend on whether the data set *A* is:

- **a population** consisting of just the numbers 31, 36, 44, 39, 40, with no other numbers involved; *or*
- **a sample** of five replicate experimental measurements, which form part of a much larger *source population* of measurements that *could be created* if the experiment were repeated many times.

We can test whether a data set is a *sample* or a *population* by considering the effect of *repeating the process by which the data values were identified*.

If repeating the process of identification still gives the same values, then the data set is a *population*, but if different values can be produced then the set is a *sample*.

A student collects **all** 200 frogs (a *population*) from a specific pond and, finding that 120 are female and 80 male, he concludes that the proportion of females in that pond is 0.60. A second student collects **all** the frogs again (the same *population*) and will record exactly the same proportion of females to males. By comparison, a student who randomly picks just 50 frogs only collects a *sample*, and may get a different proportion of females and males in every different sample of 50 frogs that he may select – see Example 14.7.

A *sample* of values only gives us *best estimates* of the *population* values that we are trying to measure. However it is possible, through good experiment design (15.1), to obtain results that are sufficiently accurate for the purpose, without the need to make an excessive number of measurements.

Q7.4

Identify, for each of the following data sets, whether the data set is a *sample* or a *population*:

(i)   The four numbers {5.1, 5.2, 5.1, 5.3}.   Sample/Population

(ii) Four repeated measurements of the pH of a   Sample/Population
      solution.

(iii) The numbers recorded in 37 spins of a roulette   Sample/Population
      wheel.

(iv) The 37 numbers (including '0') on a roulette   Sample/Population
      wheel.

(v)   The scores achieved by England in all their   Sample/Population
      football matches during the 2002 World Cup.

(vi) The amount of pocket money received by a   Sample/Population
      random selection of children as part of a
      'lifestyle' survey.

If data set A (Example 7.4) is a self-contained *population*, then we can use statistics simply to *describe* the characteristics of the data set.

If, however, the data set A is a *sample* of *experimental* values, then we will use statistics to *estimate* the characteristics of the *source population* of *experimental* measurements from which the sample was derived. For example, we will use the average (mean) value of the sample data to *estimate* what would be the average (mean) value if the experiment were repeated many more times.

A **parameter** is a *variable* that is used to describe some characteristic of a *population*. A parameter is usually given a Greek letter as a symbol, e.g. μ for population mean and σ for population standard deviation.

A **statistic** is a *variable* that is used to describe some characteristic of a *sample*, e.g. $\bar{x}$ for sample mean and s for sample standard deviation.

The value of the statistic of a sample (e.g. mean of the sample) is used to estimate the value of the equivalent parameter of the population (i.e. mean of the population) from which the sample was drawn. Different samples from the *same* population will typically give different values for the same statistic (i.e. different *sample* means).

7.2.3 Statistics notation

Typically, a calculation in statistics may be dealing with very many data values, and it would be time consuming to write out every single number. Instead, statistics often use a form of shorthand to show what is happening in the calculations.

Subscripts are used to identify *particular values* in a set of data. For example, the values from data set A in Example 7.4 could be written as a_1, a_2, a_3, a_4, a_5, where $a_1 = 31, a_2 = 36$, etc.

The *Greek symbol* $\sum$ (capital sigma) is often used to indicate the *summation* (addition) of a range of numbers: $\sum_1^5 (a_i)$ is shorthand in statistics for saying 'add together a_1, a_2, a_3, a_4, a_5'.

Example 7.5

Using the values of data set A from Example 7.4, the summation is written as:

$$\sum_1^5 (a_i) = a_1 + a_2 + a_3 + a_4 + a_5 = 31 + 36 + 44 + 39 + 40 = 190$$

In general, for n values of x to be added together, we would write:

$$\sum_1^n (x_i) = x_1 + x_2 + x_3 + \cdots + x_{n-1} + x_n \qquad [7.1]$$

The above equation says 'sum all values of x_i for i ranging from 1 to n'.

Note that when the values of *all* the possible variables in a set are to be added together, the summation is often written in simplified forms:

$$\sum_{1}^{n}(x_i) = \sum_{i}(x_i) = \sum(x_i)$$

It is well worth taking a few minutes to understand each new symbol when it appears. The ability to 'read' statistics more fluently will be an enormous help in developing a good understand of statistics as a whole.

Q7.5

Calculate $\sum_{1}^{5}(x_i)$ for the following set of *x*-values:

$$x_1 = 4.9, x_2 = 4.7, x_3 = 5.1, x_4 = 4.9, x_5 = 4.4$$

7.2.4 Means and averages

The **mean** (or **average value**) for a *population* is usually written as μ(mu).

The **mean** (or **average value**) for a *sample* is usually written as $\bar{x}$.

The mean value, $\bar{x}$, of a *sample* of data is the **best estimate** for the unknown mean, μ, of the *source population* from which the sample was derived. Note that $\bar{x}$ is a *statistic*, which is used to *estimate* the value of the *parameter, μ*.

For example, opinion pollsters take samples of about 1000 people to get best estimates for the way in which the population may choose to vote in a political election. The voting intentions of the population are *estimated* from the sample results.

The mean values for the sample, $\bar{x}$, and the population, μ, are both calculated as the simple averages of the data values:

$$\text{Sample mean, } \bar{x} = \frac{\text{Sum of all data values}}{\text{Number of data values}} = \frac{\sum_{1}^{n}(x_i)}{n} = \frac{\sum(x_i)}{n} \qquad [7.2]$$

$$\text{Population mean, } \mu = \frac{\text{Sum of all data values}}{\text{Number of data values}} = \frac{\sum_{1}^{n}(x_i)}{n} = \frac{\sum(x_i)}{n} \qquad [7.3]$$

Excel uses the function AVERAGE to calculate the mean value – see Appendix I.

Example 7.6

Calculate the *sample* mean of data set A in Example 7.4:

$$\text{Sample mean, } \bar{x} = \frac{\sum_1^5 (a_i)}{5} = \frac{31 + 36 + 44 + 39 + 40}{5} = \frac{190}{5} = 38$$

The *mean value* for a data set is useful because it gives the *location of the data* set on the value axis. We now know, for example, that data set A is spread around a 'central' value of 38.

Q7.6

Calculate the mean values, $\bar{x}$, for each of the following sets of x values:

(i) $x_1 = 4.9, x_2 = 4.7, x_3 = 5.1, x_4 = 4.9, x_5 = 4.4$
(ii) $x_1 = 0.25, x_2 = 0.27, x_3 = 0.19, x_4 = 0.22$

7.2.5 Experimental uncertainty, standard deviation and variance

Due to experimental uncertainty, any measurement may result in a deviation away from the true value, μ, and also away from the sample mean value, $\bar{x}$.

The **deviation**, d_i, of a specific data value, x_i, from the mean value, $\bar{x}$, of a set of sample data is given by:

$$d_i = x_i - \bar{x} \tag{7.4}$$

The deviation, d_i, of the data value tells us how far that value, x_i, is from the mean value (or 'centre'), $\bar{x}$, of the data set (Figure 7.6).

As a first step to developing a parameter/statistic that will give an 'average' measure of all of the *deviations*, we could try taking the simple average of all deviations. However, we soon find that this average is zero, because the *sum of all deviations* in any data set is always zero:

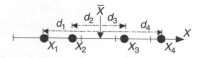

Figure 7.6 Deviation of data values from the mean.

$$\sum_1^n (d_i) = \sum_1^n (x_i - \bar{x}) = 0$$

Instead of the simple average of deviations, we can try the average of (deviations)^2, because the sum of all (deviations)^2 is *not zero*:

$$\sum_1^n (d_i^2) = \sum_1^n (x_i - \bar{x})^2 \neq 0$$

The next step is to provide an 'average' value for (deviations)^2 by dividing the sum by $n - 1$ to produce a statistic called the *sample variance* of the data:

Sample variance, $$s^2 = \frac{\sum_1^n d_i^2}{n - 1} \Rightarrow \frac{\sum_1^n (x_i - \bar{x})^2}{n - 1} \qquad [7.5]$$

However, variance cannot be used as a direct measure of *spread*, because it has units which are the *square* of the units for the data value, e.g. the variance of a measurement of time in seconds would have units of (seconds)^2.

Finally, by taking the square root of the sample variance, we produce the *sample standard deviation* (equation [7.6]), which does have the same units as the data values, and has become the most commonly used indicator to describe the *spread* of experimental data values:

Sample standard deviation, $$s = \sqrt{\frac{\sum_1^n (x_i - \bar{x})^2}{n - 1}} \qquad [7.6]$$

If we were trying to describe the *spread* of a *population*, where the mean value is given by μ, then we would use the statistic:

Population standard deviation, $$\sigma = \sqrt{\frac{\sum_1^n (x_i - \mu)^2}{n}} \qquad [7.7]$$

In most experimental situations we only have a *sample* set of data. In this case our *best estimate* of the unknown population standard deviation, σ, is given by the *sample standard deviation*, s.

The difference between the divisors in equations [7.6] and [7.7] results from the fact that σ is describing the spread of the *actual* data values, whereas s is *estimating* the spread of population values from which the sample was taken. In the latter case, $n-1$ instead of just n allows for increased uncertainty in using the sample mean, $\bar{x}$, as just a best estimate for the unknown true mean, μ.

In practice, calculations of standard deviation are usually performed either by using suitable functions in software or by using a simple hand-held calculator. In some calculators, the population standard deviation is given as σ_n and the sample standard deviation as σ_{n-1}.

It is useful to restate the relationship between variance and standard deviation:

$$\textbf{Variance} = (\textbf{Standard deviation})^2 \qquad [7.8]$$

Excel uses the functions (Appendix I): STDEV to calculate the standard deviation of a data sample; VAR to calculate the variance of a data sample; and STDEVP and VARP for the standard deviation and variance of a population.

Example 7.7

Using the data from Table 7.1 for analyst 5, calculate:

(i) Mean value, $\bar{x} = \dfrac{\sum_1^n x_i}{n}$

(ii) Sum of deviations, $\sum_1^n (d_i) = \sum_1^n (x_i - \bar{x})$

(iii) Sum of (deviations)^2, $\sum_1^n (d_i^2) = \sum_1^n (x_i - \bar{x})^2$

(iv) Sample variance, $s^2 = \dfrac{\sum_1^n d_i^2}{n-1} \Rightarrow \dfrac{\sum_1^n (x_i - \bar{x})^2}{n-1}$

(v) Sample standard deviation, $s = \sqrt{\dfrac{\sum_1^n d_i^2}{n-1}} \Rightarrow \sqrt{\dfrac{\sum_1^n (x_i - \bar{x})^2}{n-1}}$

Calculations are in the following text.

The calculation in Example 7.7 is best performed by setting up a table, as in Table 7.2.

Table 7.2. Calculation of sample standard deviation.

Data labels	Value x	Deviation $d_i = (x_i - \bar{x})$	Deviation^2 $d_i^2 = (x_i - \bar{x})^2$
x_1	81.7	0.74	0.55
x_2	83.4	0.14	0.02
x_3	78.9	1.24	1.54
x_4	75.3	0.94	0.88
x_5	79.0	−3.06	9.36

(continued overleaf)

Table 7.2. (*continued*)

Data labels	Value x	Deviation $d_i = (x_i - \bar{x})$		Deviation^2 $d_i^2 = (x_i - \bar{x})^2$
Sum, $\sum_1^n x_i =$	397.3	$\sum_1^n d_i =$	0.0	$\sum_1^n d_i^2 =$ 12.35
Sample size, $n =$	5			

$$\text{Mean, } \bar{x} = \frac{\sum_1^n x_i}{n} = \quad 79.46$$

$$\text{Sample variance, } s^2 = \frac{\sum_1^n d_i^2}{n-1} \Rightarrow \frac{\sum_1^n (x_i - \bar{x})^2}{n-1} \qquad \Rightarrow \frac{12.35}{5-1} \Rightarrow 3.09$$

$$\text{Sample standard deviation, } s = \sqrt{\frac{\sum_1^n d_i^2}{n-1}} \Rightarrow \sqrt{\frac{\sum_1^n (x_i - \bar{x})^2}{n-1}} \qquad \Rightarrow \sqrt{3.09} \Rightarrow 1.76$$

Q7.7

For the following set of x sample values:
$$x_1 = 4.9, \ x_2 = 4.7, \ x_3 = 5.1, \ x_4 = 4.9, \ x_5 = 4.4$$

(i) Calculate the mean value of the sample: $\bar{x}$.
(ii) Calculate each of the deviations: d_1, d_2, d_3, d_4, d_5.
(iii) Calculate the sum of all deviations: $\sum_1^n (d_i)$.
(iv) Will the answer for (iii) always be the same for any set of data?
(v) Calculate the sum of all (deviations)^2 : $\sum_1^n (d_i^2)$.
(vi) Calculate the sample variance: s^2.
(vii) Calculate the sample standard deviation: s.

Q7.8

For each of the three (sample) sets of experimental data values below, calculate the values in the table below and answer (iv):

	Mean	Sample variance	Sample standard deviation
(i) 8, 6, 7, 4, 7			
(ii) 68, 66, 67, 64, 67			
(iii) 418, 416, 417, 414, 417			

(iv) What similarities are there about the answers for (i), (ii) and (iii)?

7.2.6 Sample and population standard deviations

When measuring an unknown quantity, it is useful to understand the difference between the standard deviation, s, of a *sample of measurements* and the *population* standard deviation, σ, of *all the measurements* that could be made.

The sample standard deviation, s, is the best estimate, based on the sample data, for the true population standard deviation, σ.

When the sample size, n, is small, then there are likely to be considerable variations in the measured value of s, but if the sample size is increased, s will approach the true value, σ.

In Table 7.3 we have taken data from Example 7.2. The first five columns give the sample standard deviations recorded by analysts 1 to 5 in Example 7.2. These all had a sample size $n = 5$, and show a variation in the recorded sample standard deviation. The next four columns show the effect of combining results to give larger sample sizes – grouping the data from sets 1 and 2, then sets 1 to 4, 1 to 8 and finally 1 to 16, to give the standard deviations for samples of increasing size.

Table 7.3.   Sample standard deviation and sample size.

Data source	1	2	3	4	5	1+2	1–4	1–8	1–16
Sample size, n	5	5	5	5	5	10	20	40	80
Sample standard deviation, s	1.76	2.42	2.27	1.45	3.09	2.02	1.89	1.91	1.98

Example 7.8

On the basis of the results in Table 7.3 give the best estimate that can be made for the true population standard deviation for the measurement of blood–alcohol made by the analysts in Example 7.2.

The standard deviations of the individual analysts (1 to 5) show a considerable variation between 1.45 and 3.09, but as the sample size increases, the values become closer, leading to 1.98 for a sample size of $n = 80$ measurements.

Hence the best estimate available from the data in Table 7.3 is $s = 1.98$.

In fact, we know that the example data was created randomly from a true distribution with a population standard deviation $\sigma = 2.00$. The 'experimental' observations agree well with the true value, provided that the sample size is sufficiently large.

7.3 Frequency Statistics

7.3.1 Introduction

In statistics, we often need to count the number of times that a particular outcome, x, occurs, e.g. the number of *large* eggs laid by a group of chickens, or the number of students who achieve an upper second classification in their degree.

The number of times a particular outcome, x, occurs is called 'frequency'. The value of using the concept 'frequency' is that it provides a useful method of describing and handling large quantities of data. In this unit we will write the frequency of x as $f(x)$ or f_x. The two expressions are equivalent, but it is sometimes easier to print one than the other.

We will also establish the link between the frequency, $f(x)$, with which particular events have occurred in the past, with the probability, $p(x)$, with which similar events may be expected to occur in the future.

7.3.2 Plotting data values

An important function of statistics is to handle large numbers of data values in such a way that the analyst can easily understand the essential characteristics of the data set.

A **stem and leaf diagram** is a quick way of displaying a spread of data values for initial analysis.

Example 7.9

The stem and leaf diagram in Figure 7.7 shows the examination marks, M, for 50 final-year students:

```
30 | 7 9 9
40 | 1 2 2 4 5 6 7 7 8 9
(50) 1 1 1 1 1 2 2 3 3 4 5 5 5 6 8 9 9 9
60 | 0 0 1 2 2 2 3 4 4 5 5 5 6 7
70 | 0 0 1 2 7
```

Figure 7.7   Stem and leaf diagram.

The first row shows that there were students with marks 37, 39 and 39 respectively, and, similarly, the final row shows that students had marks 70, 70, 71, 72 and 77.

The 'stem' is the column on the left, which shows a value that is common to all data values on that level. The 'leaves' are the individual marks that must be added to the stem level to get the actual data value. For example, the circled bold italic values show that one student achieved a mark of $54 = 50$ (stem) plus 4 (leaf).

The value of a stem and leaf diagram is that it can be hand-drawn very quickly to review the spread of results as they are being recorded in an experiment.

Another quick way of displaying information is the **dot plot**, which just gives a 'dot' for each data value.

Example 7.10

The graph in Figure 7.8 shows a dot plot for the examination marks data for the 50 students given in Figure 7.7:

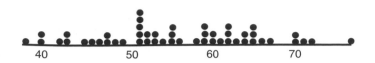

Figure 7.8   Dot plot.

Q7.9

The following data gives the weights, w (in grams), of a collection of 24 seeds.

48	46	46	66	55	55	47	44
49	40	33	57	42	25	58	32
49	51	39	67	61	42	62	27

(i) Represent the data by drawing a stem and leaf diagram.
(ii) Use the diagram to count the numbers of seeds that fall within the following ranges:

$20 \leqslant w < 30;\ 30 \leqslant w < 40;\ 40 \leqslant w < 50;\ 50 \leqslant w < 60;\ 60 \leqslant w < 70$

(Note that $20 \leqslant w < 30$ means that w is a number that is greater than *or equal to* 20, but also less than 30.)

7.3.3   Frequency data

When we have a large number of data values it can be unnecessarily cumbersome to treat them as *individual* data items. It is more convenient to divide the data range into a number of divisions, and then simply count how many data values fall in each division.

Each division, g, is called a **class** (or **bin**), and the number of data values in the class is called its **class frequency**, f_g.

For example, Table 7.4 summarizes the examination marks, M, of the 50 final-year students given in Example 7.9. The class frequency is found by counting the number of values in each class.

The degree grade (class) is decided on the basis of the *numerical range* within which the student's aggregate mark, M, falls.

Each class may also be described as a *category* and given a **categorical name**: Fail (F), Third (3rd), Lower Second (2.2), Upper Second (2.1) and First (1st).

Table 7.4. Final-year performance of 50 students.

Mark range	Categorical name	Class frequency		Relative frequency
$M < 40$	Fail (F)	$f(F) =$	3	$3/50 = 0.06$
$40 \leqslant M < 50$	Third (3rd)	$f(3rd) =$	10	$10/50 = 0.20$
$50 \leqslant M < 60$	Lower Second (2.2)	$f(2.2) =$	18	$18/50 = 0.36$
$60 \leqslant M < 70$	Upper Second (2.1)	$f(2.1) =$	14	$14/50 = 0.28$
$M \geqslant 70$	First (1st)	$f(1^{st}) =$	5	$5/50 = 0.10$
		Total $=$	50	1.00

The results are expressed as the number, class frequency, $f(g)$ or f_g, of students achieving each grade (class, g) of degree: e.g. three students fell into the 'Fail' category and five students were in the 'First' category.

The *proportion* of data values in each class is given by the **relative class frequency** (see Table 7.4):

$$\textbf{Relative class frequency} = \frac{\textbf{Class frequency}}{\textbf{Total of all frequencies}} \qquad [7.9]$$

Relative class frequency may often be expressed as a percentage.

Q7.10

A group of 24 graduates were surveyed to record their salaries, S, two years after graduation. The results, measured in £000s, were as follows:

20.1	17.7	22.2	15.0	22.0	23.7	15.4	16.7
14.7	18.7	20.4	24.5	26.2	24.9	17.7	21.3
16.2	21.7	17.1	17.0	19.0	12.0	18.8	18.5

Use the table below to record answers to the following questions:

(i) Count the number (frequency) of graduates, $f(g)$, who achieve **salaries, S,** in the ranges (classes), g, that are given in the table below.
(ii) Calculate the relative frequency (rounding to 3 significant figures) for each of the classes, g.
(iii) Calculate the 'total' for all relative frequencies.
(iv) Is the answer in (iii) what was expected?

Salary range	Number (frequency)	Relative frequency
g	$f(g)$	
$12.0 \leqslant S < 16.0$		
$16.0 \leqslant S < 20.0$		
$20.0 \leqslant S < 24.0$		
$24.0 \leqslant S < 28.0$		
Totals:	24	

7.3.4 Plotting frequency data

If the data values are grouped into different *categories*, we can conveniently plot this **categorical** class data using a column graph or a pie chart. For example, Figure 7.9 is a **column graph (or bar graph)** which records the performance (Table 7.4) of the cohort of 50 final-year students (from Example 7.9).

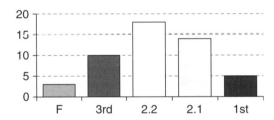

Figure 7.9   Column (or bar) graph for the final-year performance of 50 students.

Note that Excel draws a column graph with *vertical* 'columns' and a bar graph with *horizontal* 'bars'.

The overall shape of the column (or bar) graph conveys a lot of information, without the need to write down the mark of every individual student.

In a column graph, each class frequency is proportional to the **height** of the relevant column.

Figure 7.10 is a **pie chart** that records the same data as Figure 7.9, but the values for each class are recorded as *percentage relative frequencies*.

In a pie chart, each class frequency is proportional to the **angle** of the relevant slice.

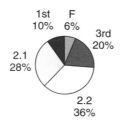

Figure 7.10 Pie chart for the final-year performance of 50 students.

Q7.11

The following data gives the areas of four regions of the United Kingdom, given in units of 1000 km^2.

	Area (1000 km^2)	Fraction	Angle (degrees)
England	130	0.533	192
Northern Ireland	14		
Scotland (incl. Islands)	79		
Wales	21		
Total:	244		

(i) Represent the area data by drawing a column graph.
(ii) Calculate the fraction of the total area covered by each region, by dividing the area of the region by the total, e.g. England covers a fraction 130/244 = 0.533
(iii) Multiply each fraction by 360 to calculate the angles for a pie chart, e.g. for England 0.533 × 360 = 192°.
(iv) Plot the data as a pie chart.

7.3.5 Histograms

The data can be plotted as a histogram, provided that a *quantitative variable* can be plotted along the x-axis instead of *categorical data*, e.g. plotting the actual degree marks instead of the degree classifications – see Figure 7.11.

It is very important to understand that, in a **histogram**:

- The **area of each block** (= height × width) is proportional to the *class frequency*.

- The data is *continuous* along the value axis, i.e. there are no *gaps between the blocks* (unless a block has a zero height).
- The *total area of a frequency histogram* is equal to the *total number*, n, of data values.

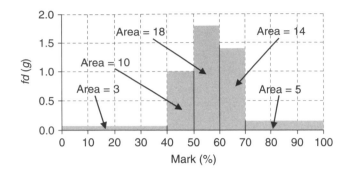

Figure 7.11   Histogram for the final year performance of 50 students.

The *height* of each block on the ordinate (y-axis) scale of the histogram is given by the **frequency** *density, fd (g)*:

$$\text{Frequency density, } fd(g) \text{ or } fd_g = \frac{\text{Class frequency}}{\text{Width of class}} = \frac{f_g}{\Delta x_g} \qquad [7.10]$$

The Greek symbol Δ is often used to indicate a difference in a variable – see 6.1.2. In the above equation Δx refers to a small change in the x variable.

The **area** of each block is given by **class** *frequency, f (g)*, which, by rearranging [7.10], is given by:

$$f(g) \text{ or } f_g = fd(g) \times \Delta x_g \qquad [7.11]$$

Example 7.11

Figure 7.11 shows the **histogram** for the final-year performance of 50 students (from Figure 7.7).

The 'heights' (frequency density) of the histogram blocks were calculated as follows:

Class	Class frequency	Class range	Class width	Frequency density
g	$f_g = f(g)$		$\Delta x_g = \Delta x(g)$	$fd_g = fd(g)$
Fail	3	$0 \leqslant x < 40$	40	$= 3/40 = 0.075$
Third	10	$40 \leqslant x < 50$	10	$= 10/10 = 1.0$
2.2	18	$50 \leqslant x < 60$	10	$= 18/10 = 1.8$
2.1	14	$60 \leqslant x < 70$	10	$= 14/10 = 1.4$
First	5	$70 \leqslant x \leqslant 100$	30	$= 5/30 = 0.167$

Compare the shape of the column graph (drawn using categories) in Figure 7.9 with the histogram (drawn using a continuous quantitative axis) in Figure 7.11. The differences occur when the *class widths are not the same*, i.e. for the 'Fail' and 'First' categories.

Q7.12

In Year 8 of school, 84 children were given a project to record the distribution of their heights, h (cm); they chose to count the number of pupils whose heights fell within specific ranges. However, they used different widths for the different ranges, with the result that the final data is recorded as follows:

Class range	Class frequency	Class width	Frequency density
h (cm)	$f_g = f(g)$	$\Delta x_g = \Delta x(g)$	$fd_g = fd(g)$
$144 < h \leqslant 154$	6	10	$= 6/10 = 0.6$
$154 < h \leqslant 158$	21	4	$= 21/4 = 5.25$
$158 < h \leqslant 160$	17		
$160 < h \leqslant 162$	15		
$162 < h \leqslant 166$	16		
$166 < h \leqslant 176$	9		

(i) Use the empty columns of the table to calculate the class width and class density $(fd(g) = f(g)/\Delta x(g))$ for each class (two classes have already been completed).

(ii) Decide if it is possible to present the data graphically using a histogram (even though the class widths are not all the same).

(iii) If the answer to (ii) is Yes, then draw the relevant histogram.

7.3.6 Mean and standard deviation

In frequency data we often do not know the exact value of *each* data item. We only know that its value falls somewhere within a particular class range, e.g. a student with an upper second grade has a mark somewhere between 50 and 59.

We make the *approximation* that *every data item* in a particular class, g, has a value equal to the *mean class value*, $\bar{x}_g$, of that class.

The mean class value is obtained by calculating the average of all the data values that *could be* included in that class. For example, the mean class value of a class that has *any value* in the range $50 \leqslant x < 60$ would be $\bar{x}_g = 55$.

Note, however, that the mean class value of a class that has *only integer values* 50 to 59 would be the mean value of 50, 51, 52, 53, 54, 55, 56, 57, 58 and 59, giving $\bar{x}_g = 54.5$.

Thus:

Approximate sum of the f_g **values** in **class**, $g = f_g \times \bar{x}_g$ [7.12]

For example, if the mean class value is $\bar{x}_g = 55$, and there are $f_g = 4$ items in that class (e.g. 51, 53, 55, 58), then the *approximate* sum of those four values is $f_g \times \bar{x}_g = 4 \times 55 = 220$. This is not the exact sum ($= 217$ for the example values) but will often give a sufficiently good approximation.

In the equations below, the symbol, $\sum_g$, means 'add up the following expression for all classes'. For example, $\sum_g (f_g)$ means 'add up the values of f_g for all classes':

Total number of data values : $n = \sum_g (f_g)$ [7.13]

Approximate sum of *all* values in *all* classes: $\text{Sum} = \sum_g (f_g \times \bar{x}_g)$ [7.14]

We can derive approximate equations for calculating the mean and standard deviation.
Sample mean values:

$$\bar{x} = \frac{\sum_g (f_g \times \bar{x}_g)}{n}$$ [7.15]

Sample standard deviations:

$$s = \sqrt{\frac{\sum_g \left[f_g \times (\bar{x}_g - \bar{x})^2 \right]}{n - 1}}$$ [7.16]

Example 7.12

The number of students in a sample group obtaining (integer) marks in different ranges is given in the table below. Use Excel to calculate the mean and sample standard deviation for the distribution of frequency values.

Range	0–9	10–19	20–29	30–39	40–49	50–59	60–69	70–79	80–89	90+
Number	0	1	4	18	39	80	47	9	2	0

Note (for example) that the mean value, x_g, for the range 30–39 will be 34.5 when the possible scores have only integer values. A similar calculation applies for all ranges.

Using [7.15] and [7.16], a calculation in Excel gives:

$$\text{Mean value, } \overline{x} = 53.50$$

$$\text{Standard deviation, } s = 11.47$$

Q7.13

In this question we compare *two ways* of calculating the *sum* and the *mean* of a set of data. We will use the same data for the salaries, S, of 24 graduates as used in Q7.10.

20.1	17.7	22.2	15.0	22.0	23.7	15.4	16.7
14.7	18.7	20.4	24.5	26.2	24.9	17.7	21.3
16.2	21.7	17.1	17.0	19.0	12.0	18.8	18.5

In the first part, we estimate (iv) the sum and (v) the mean by first dividing the data values *into classes* and using the mean class values.

We then compare these values to (vi) the sum and (vii) the mean calculated by using each *individual* data value.

(i) Using classes g, as in the table below, count the frequency, f_g, for each class, and the total, n, for all classes. (Check these with the answer to Q7.10.)
(ii) Calculate the mean class values, $\overline{S}_g$, for each class, g. (Take the mean of all possible values within each range, e.g. $\overline{S}_g$ for the first class is 14.0.)
(iii) For each class g, multiply the number (f_g) in the class by the mean class value, $\overline{S}_g$, to give the class sum, $f_g \times \overline{S}_g$, and enter the value in the right-hand

column. This is an *estimate* of the sum of salaries for all graduates in that class.

(iv) Find the sum, $\sum_g f_g \times \overline{S}_g$, for all classes. This gives an estimate of the *sum of all 24 graduates' salaries*.

(v) Estimate the *mean value*, $\overline{S}$, of *all* the salaries by dividing the value in (iv) by the number of graduates, n.

(vi) Find the true sum by adding all the *individual* salaries. Is this value the same as the result in (iv)?

(vii) Find the true mean by dividing the value in (vi) by the number of graduates, n. Is this value the same as the result in (iv)?

(viii) When might the method used in steps (i) to (v) be preferable to the more accurate method used in (vi) and (vii)?

Salary range	Frequency	Mean class value	Class sum
Class, g	f_g	$\overline{S}_g$	$f_g \times \overline{S}_g$
$12.0 \leqslant S < 16.0$		14.0	
$16.0 \leqslant S < 20.0$			
$20.0 \leqslant S < 24.0$			
$24.0 \leqslant S < 28.0$			
$n = \sum_g (f_g) =$		Sum $= \sum_g (f_g \times \overline{S}_g) =$	
	Mean value, $\overline{S} = \dfrac{\sum_{\cdot g} (f_g \times \overline{S}_g)}{n} =$		

7.3.7 Relative frequency and probability

The outcomes of random events (or 'trials') are often governed by particular probabilities, e.g. the scores recorded by six-sided cubic dice are governed by a *1 in 6* chance.

Frequency data, $f(g)$, can be used to describe the number of times particular *outcomes*, g, *have occurred*.

Probability, $p(g)$, is a *prediction* of the relative frequency with which *future outcomes*, g, can be *expected to occur*.

The **relative frequency** with which particular events *have occurred* in the *past* can sometimes be used to predict the *probability* with which similar events *will occur* in the *future*:

$$p(g) = \frac{f(g)}{n} \qquad\qquad [7.17]$$

A **frequency distribution** is easily transformed into a **probability distribution**.

The *area* under a *frequency* distribution equals n, the sum of the individual frequencies.

The *area* under a *probability* distribution equals 1.0, the sum of the individual probabilities.

Q7.14

In a game where two six-sided dice are thrown (e.g. backgammon, Monopoly), the combined score, x, can be any value from 2 to 12.

(i) Work out the number of ways, $f(x)$, of obtaining each score value x. For example, there are two ways of getting a combined score of 3 using two dice: $1 + 2$ or $2 + 1$, giving $f(3) = 2$.

(ii) Present the results as a histogram on the chart, Figure 7.12 (in this case, frequency density, $fd(x)$, equals frequency, $f(x)$, because the class width is equal to 1).

(iii) What is the total 'area' of the histogram?

(iv) What does the total 'area' of the histogram represent?

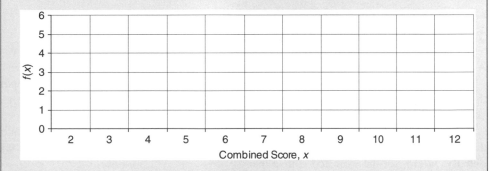

Figure 7.12

Q7.15

Convert the 'frequency' histogram, $f(x)$, produced in Q7.14 into a 'probability' histogram, $p(x)$. Remember that, when rolling two dice, there are a total of 36 different possible outcomes.

(i) For each combined score, x, calculate the probability, $p(x)$, that one roll of the two dice will give that particular result. For example, there are two ways of getting a combined score of 3 using two dice: $1 + 2$ or $2 + 1$. There are a total of 36 different possible outcomes. Hence the probability of getting a combined score of 3 is $p(3) = 2/36 = 0.0556$.

(ii) Present the results as a histogram on the chart, Figure 7.13.
(iii) What is the total 'area' of the histogram?
(iv) What does the total 'area' of the histogram represent?

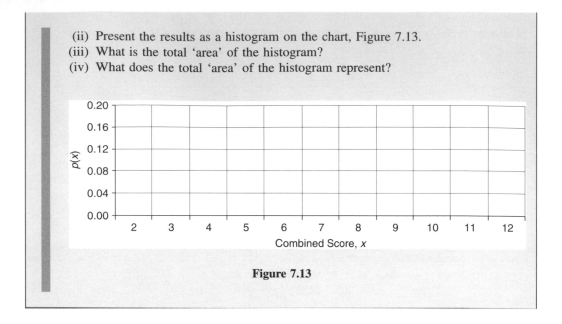

Figure 7.13

7.3.8 Continuous distribution

A histogram is normally characterized by a number of clearly visible blocks, e.g. Figure 7.11.

If the *width* of each block is *reduced* and the *number* of blocks *increased* in proportion, then the envelope (outer curve) of the histogram will generally become 'smoother', i.e. we change from a histogram to a *continuous* distribution.

For example, on the basis of past student performance, a particular university estimates the distribution of students who will gain different course marks. Figure 7.14 shows the **probability density**, $pd(x)$, for the expected course marks.

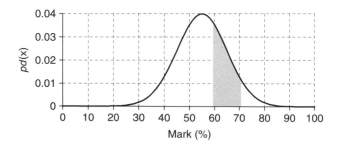

Figure 7.14   Probability density of course marks.

In a continuous probability distribution, the total area under the curve is 1.0, i.e. any student who completes the course has a mark that falls *somewhere* in the range covered by the curve. For a *continuous* distribution, we assume that the course mark can be calculated to any specific value and not just presented as an integer.

In a continuous probability distribution, the *probability* that a trial falls within a given range of x-values is given by the *area under the curve* between those values.

Example 7.13

In Figure 7.14, the probability that a student, selected at random, will have a mark between 60% and 70% is equal to the shaded area under the curve. We can make a rough estimate of this probability by noting that:

- the width of the area $= 70 - 60 = 10$; and
- the average height of the area is about 0.025.

This gives a probability area of about $0.025 \times 10 = 0.25$, i.e. there is about a 25% chance that any particular student will get a mark in this range (an upper second). An exact calculation gives a probability of 24.17%.

Q7.16

Use the graph in Figure 7.14 to make a **rough** estimate of the probability that a student, selected at random, will have a course mark between 30% and 40%.

7.3.9 Cumulative Probability

Cumulative probability, $cp(x)$, is the probability that the result of a trial may fall anywhere in the range from $-\infty$ to x.

The cumulative probability, $cp(x)$, for a specific value of x is equal to the probability area under the $pd(x)$ curve from $-\infty$ to x.

In situations where the value of x can only be *positive* (as for examination marks), the cumulative probability, $cp(x)$, is equal to the probability area from 0 to x.

The cumulative probability, $cp(x)$, will increase from 0 to 1.0 as the value of x increases.

Using the example from Figure 7.15, the probability ($= 0.84$) of selecting a student with a mark between 0% and 65% is given by the shaded area in Figure 7.15. This probability is

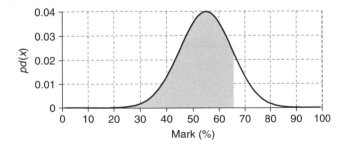

Figure 7.15 Probability area (shaded) for a mark between 0% and 65%.

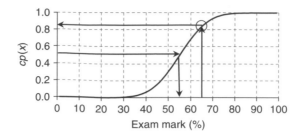

Figure 7.16   Cumulative probability of exam scores.

represented by the circled point on the cumulative probability curve in Figure 7.16: 0.84 of students will get a mark of 65 % or less and so $1 - 0.84 = 0.16$ of students will get a mark greater than 65 %.

Similarly, the cumulative probability curve in Figure 7.16 shows that half (probability = 0.5) of all students will get a mark of 55 % or less, and all students (probability = 1.0) will get 100 % or less!

Q7.17

Using the data from the 'probability' histogram produced in Q7.15:

(i) For each combined score, x, calculate the cumulative probability that one roll of the two dice will give that score or any score below that value.
(ii) What is the cumulative probability that a roll of two dice will give a combined score of 6 or less?
(iii) What is the probability that a roll of two dice will give a combined score of at least 10?

7.4 Probability

7.4.1 Introduction

Probability is a measure of the *certainty* with which we may expect a particular outcome to occur. Probability is based on a scale of 0 to 1.

An *impossible* event, e.g. a cow jumping over the Moon, has a probability of 0: p(cow will jump over Moon) $= 0.0$.

A *certain* event, e.g. an iron bar sinking in water, has a probability of 1: p(iron bar will sink in water) $= 1.0$.

The concept of probability is often used when the outcome of some event (or trial) *appears* to be governed solely by chance, e.g. in rolling dice in a board game.

The fact that many outcomes *appear* to be governed by chance is often due to the complexity of the factors leading up to the particular outcome. The behaviour of playing dice, as they bounce and roll on the table, is actually governed by the laws of mechanics. However, the complexity of their motion is such that it is impossible to control their behaviour and predict the number that will be shown. In practice, the numbers shown by dice are *random* and unpredictable – each number has an equal chance of appearing (assuming that the dice are not biased!).

Excel (Appendix I) can be used for the logical operations AND, OR, NOT and IF.

7.4.2 Probability and frequency

There are two main ways of estimating the probability with which a future event may occur:

- *Calculated on the basis of the science* of the situation, e.g. as playing dice are symmetrical cubes, we would expect an *equal* probability for any number to appear.
- *Estimated from the frequency* with which particular events have *occurred in the past*, e.g. past experience may show that one in three eggs laid by a particular chicken are small eggs and we may then expect that the future probability would also be one in three.

Frequency, in statistics, is a measure of the number of times a particular outcome occurs: $f(X)$ represents the number of times the outcome, X, occurs.

If we group the possible outcomes into different *classes*, then the *class frequency* is the number of times an outcome falls into a particular class (see also 7.3.3).

Example 7.14

For example, if we collect 180 eggs and divide them into 'small', 'medium' or 'large' categories (or classes), we may find that we have 60 'small' eggs, 80 'medium' eggs and 40 'large' eggs. The class frequencies in this case are described by the frequencies, f(small), f(medium), and f(large) respectively. This particular situation is described in Table 7.5.

Table 7.5. Example of a frequency (or probability) distribution of egg sizes.

Class	Class frequency		Relative frequency		Probability		
Small	f(small) =	60	f(small)/n =	60/180	p(small)	= 1/3	≈ 0.333
Medium	f(medium) =	80	f(medium)/n =	80/180	p(medium)	= 4/9	≈ 0.445
Large	f(large) =	40	f(large)/n =	40/180	p(large)	= 2/9	≈ 0.222
Totals:	n =	180				= 1/3+4/9+2/9 = 1	= 1.000

Relative frequency [7.8] is the observed frequency *divided* by the total, n, of all frequencies.

The sum of all relative frequencies is 1.00.

In the 'egg' example in Example 7.14, there were 180 'events' (sometimes called 'trials' in statistics), and, for each event or trial, there were three possible class 'outcomes': small, medium or large.

The **probability**, $p(X)$, of a future outcome, X, is a *prediction* of the *relative frequency* of that outcome *assuming the same situation could be repeated many times in the future*:

$$p(X) = \frac{f(X)}{n} \qquad\qquad [7.18]$$

Similarly, the future **frequency**, $f(X)$, of a particular outcome, X, can be calculated by multiplying the probability of that particular outcome, $p(X)$, by the number, n, of events or trials:

$$f(X) = p(X) \times n \qquad\qquad [7.19]$$

Taking the example of the eggs in Example 7.14, we know that the relative frequency, $f(\text{small})/n$, of getting a 'small' egg was 1/3, based on 180 previous eggs. We can then make a prediction for the probability that the *next* egg will be small, $p(\text{small}) = 1/3$.

Similarly we can also predict the other probabilities, $p(\text{medium}) = 4/9$ and $p(\text{large}) = 2/9$ – see Table 7.5.

When we predict *future probabilities* based on *past frequencies*, we are assuming that there will be no changes that may affect the outcomes, e.g. a change in diet for the chickens may reduce the proportion of small eggs.

A probability calculated on the basis of past experience is called an *experimental* probability.

Q7.18

Over a number of years it is found that the 140 graduates of the honours degree in horticulture include 34 with a third-class degree, 50 with a lower second degree, 38 with an upper second degree and 18 with a first-class degree.

(i) Calculate the *relative frequencies* for each class of degree.
(ii) Estimate the future *probability* that a particular student may gain a first-class degree.
(iii) What assumptions were made in order to answer (ii) above?
(iv) Estimate the number of students, in a cohort of 30 graduates, who would attain each of the four classes of degree.

In many other cases, we can use the science of the problem to *calculate* the relative frequencies without the need to carry out extensive trials. For example, we can use the *symmetry* of six-sided dice to calculate that the relative frequency (and hence probability) of getting any specific number (e.g. a 5) on a normal die. Each 'number' has an equal chance (1 in 6) of occurring:

$$p(1) = p(2) = p(3) = p(4) = p(5) = p(6) = 1/6$$

where $p(1)$, $p(2)$, etc., are the probabilities of recording a 1, 2, etc., when throwing the die.

A probability that is *calculated* using the science and mathematics of the problem is called a *theoretical* probability.

Example 7.15

The numbers observed of gene types AB, Ab, aB, ab are expected to be in the ratio 9:3:3:1 respectively, and each observation will give one of these four.

(i) Calculate the *probabilities* of observing each of the four types.
Starting with the ratios, we know that, in a total of $n = 9 + 3 + 3 + 1 = 16$ observations, we would expect to see 9, 3, 3, 1 of each type respectively.
Using the equation $p(X) = f(X)/n$, we can calculate the future probabilities from the observed frequencies:

$$p(AB) = 9/16, \ p(Ab) = 3/16, \ p(aB) = 3/16, \ p(ab) = 1/16$$

(ii) Calculate the *number* of each type that would be expected out of a total of 200 observations.
Out of a new total of 200 observations, we can calculate the expected number for each type by using [7.19]: $f(X) = p(X) \times n$ giving

$9/16 \times 200 = 112.5$

$3/16 \times 200 = 37.5$

$3/16 \times 200 = 37.5$

$1/16 \times 200 = 12.5$

Q7.19

A standard pack of 52 playing cards contains four suits (hearts, clubs, diamonds and spades), and each suit contains one each of cards numbered 2 to 10, plus a jack, a queen, a king and an ace. The jack, queen, king are called picture cards.

Calculate the probabilities of selecting at random:

(i) the king of clubs   (iv) an ace
(ii) a king (of any suit)   (v) a card of the club suit
(iii) a picture card

7.4.3 Combining independent probabilities with AND

If two *independent* trials occur, with the probability of getting A in the first trial being $p(A)$ and the probability of getting B in the second trial being $p(B)$, then the probability of getting A followed by B is:

$$p(A \text{ AND } B) = p(A) \times p(B) \qquad [7.20]$$

Note the use of '$\times$' with the 'AND' combination.

Example 7.16

Calculate, using the egg example from Table 7.5, the probability, when picking two eggs, that the first egg is small and the second is large:

$$p(\text{small AND large}) = p(\text{small}) \times p(\text{large}) \Rightarrow (1/3) \times (2/9) \Rightarrow 2/27$$

Note that the multiplication of probabilities for the AND combination assumes that the two *probabilities are independent of one another*. The outcome for the first event or trial must have no influence over the outcome for the second event.

Q7.20

It is observed that in a particular group of plants: 75 % are tall and 25 % are short; and 75 % have a round seed shape and 25 % have a wrinkled seed shape.

Calculate the probability of finding each of the following four combinations (assume that the height and seed shape are not related):

(i)   tall with a round seed
(ii) short with a round seed
(iii) tall with a wrinkled seed
(iv) short with a wrinkled seed

7.4.4 Combining probabilities with OR

If A and B are two of the possible outcomes of a trial, which have probabilities of $p(A)$ and $p(B)$ respectively, then the probability of *either A or B* occurring is given by:

$$p(A \text{ OR } B) = p(A) + p(B) - p(A \text{ AND } B) \qquad [7.21]$$

Note the use of '+' for the 'OR' combination.

If the outcomes A and B are *mutually exclusive*, i.e. they cannot occur together and $p(A \text{ AND } B) = 0$, then we have the simple form of the equation:

$$p(A \text{ OR } B) = p(A) + p(B)$$

Example 7.17

(i) Calculate, using the data from Table 7.5, the probability of getting *either* a medium egg *or* a large egg:

$p(\text{medium OR large}) = p(\text{medium}) + p(\text{large}) \Rightarrow 4/9 + 2/9 \Rightarrow 6/9 \Rightarrow 2/3$

(ii) Calculate the probability of getting a small egg *or* a medium egg *or* a large egg:

$p(\text{small OR medium OR large}) = p(\text{small}) + p(\text{medium}) + p(\text{large})$

$\Rightarrow 1/3 + 4/9 + 2/9 \Rightarrow 1$

The probability of '1' in (ii) shows that it is certain ($p = 1$) that the egg will definitely be one of the three possible sizes!

Q7.21

A trial has four possible outcomes, A, B, C and D, which have the following probabilities:

$$p(A) = 1/8, \ p(B) = 1/6, \ p(C) = 1/3, \ p(D) = 9/24$$

Calculate the probabilities that a single outcome will result in:

(i) A or B (i.e. calculate the probability $p(A \text{ OR } B)$)
(ii) A or B or C
(iii) A or B or C or D

Q7.22

Using the standard pack of playing cards described in Q7.19, calculate the probability of selecting at random:

(i) the king of clubs *or* the king of diamonds,
(ii) the king of hearts *or* the king of clubs *or* the king of diamonds *or* the king of spades,
(iii) any picture card *or* any ace.

7.4.5 Combining probabilities with NOT

If A is one possible outcome of a trial, then the outcome 'NOT A' means that the outcome A has *not* occurred.

'NOT A' is normally written as $\overline{A}$, and sometimes (for convenience on web pages) as $\sim A$. Do not confuse $\overline{A}$ (NOT A) with $\overline{x}$, which is the mean value of a set of x data!

If an event or trial has possible outcomes A, B, C, then the 'NOT A' condition requires that either B or C must occur:

$$p(\overline{A}) = p(B \text{ OR } C) = p(B) + \mathrm{p}(C)$$

In any event or trial where A is one possible outcome, then *either A or* NOT A *must* definitely occur, hence:

$$p(A \text{ OR NOT } A) = 1$$

which is the same as:

$$p(A) + p(\overline{A}) = 1$$

giving:

$$p(\overline{A}) = 1 - p(A) \qquad [7.22]$$

Example 7.18

Calculate, using the data from Table 7.5, the probability of getting an egg that is NOT small.

In the case of three categories of eggs, the outcome of 'NOT small' means that the egg produced must be either 'medium' or 'large'.

Using the data from Table 7.5:

$$p(\text{NOT small}) = 1 - p(\text{small}) \Rightarrow 1 - 1/3 \Rightarrow 2/3$$

which gives the same result as using:

$$p(\text{NOT small}) = p(\text{medium OR large}) \Rightarrow 4/9 + 2/9 \Rightarrow 6/9 \Rightarrow 2/3$$

Q7.23

Using the standard pack of playing cards described in Q7.19, calculate the probability of selecting at random:

(i) a king (of any suit),
(ii) a card that is not a king,
(iii) either a king or a card that is not a king.

7.4.6   Multiple outcomes

For more complicated probability outcomes, it is often possible that one *overall outcome* may be achieved by any one of several different separate outcomes. In this case, the overall probability is given by the *addition of the probabilities of the separate outcomes*.

Example 7.19

Calculate the probability, when two dice are thrown together, of getting a total score of 4.

When two dice are thrown together, the important *overall outcome* is the *sum* of the two dice values. The same overall *sum* can be achieved (except for 2 and 12) in more than one way, e.g. an overall value of 4 can be achieved from the values of the two dice in three ways:

$$1 + 3 \text{ or } 2 + 2 \text{ or } 3 + 1$$

(Note that the outcome $1 + 3$ is different from $3 + 1$ – swapping the values on the two dice.)

The overall probability of getting a *total* of 4 from two dice is therefore:

$$p(\text{sum of } 4) = p(1 \text{ AND } 3) + p(2 \text{ AND } 2) + p(3 \text{ AND } 1)$$

The probability of throwing a 1 followed by a 3 is given by:

$$p(1 \text{ AND } 3) = p(1) \times p(3) \Rightarrow 1/6 \times 1/6 \Rightarrow 1/36$$

Similarly: $p(2 \text{ AND } 2) = p(3 \text{ AND } 1) = 1/36$ giving:

$$p(\text{sum of } 4) = 1/36 + 1/36 + 1/36 \Rightarrow 3/36 \Rightarrow 1/12$$

Q7.24

When rolling two dice, calculate the probabilities of scoring a *total* of exactly:

(i) 2   (iii) 7
(ii) 3   (iv) 12

Q7.25

On average, 90 % of seeds of a given type of seed have been found to germinate successfully.

Calculate the probabilities that:

 (i) one selected seed will germinate;
 (ii) one selected seed will not germinate;
 (iii) every one of four selected seeds will germinate;
 (iv) none of four selected seeds will germinate;
 (v) just one of four selected seeds will germinate (hint: consider how many different ways it is possible to have one seed germinating and three seeds not germinating);
 (vi) three of four selected seeds will germinate.

7.4.7   Conditional probability

The probability of a given outcome (e.g. getting wet, W) may be dependent on (*conditional* on) an existing condition (e.g. it is raining, R). Such a conditional probability would be written as $p(W|R)$, i.e. the probability of W occurring, given that condition R exists.

Quite clearly, the probability of getting wet if it is *not* raining, $p(W|\overline{R})$, will generally be considerably less than the probability of getting wet if it *is* raining, $p(W|R)$.

A good example of conditional probability occurs with diagnostic testing for a particular disease.

The probability of getting a positive (P) result from a diagnostic test for a particular illness will normally depend on whether the person being tested has (D), or does not have ($\overline{D}$), the disease. Note that a *positive* result for the test indicates that the person *does* have the disease.

The probability of a:

- *true* positive result (having the disease) would be given by $p(P|D)$;
- *false* positive result (not having the disease) would be given by $p(P|\overline{D})$;
- *true* negative result (not having the disease) would be given by $p(\overline{P}|\overline{D})$;
- *false* negative result (having the disease) would be given by) $p(\overline{P}|D)$.

The use of conditional probability can be illustrated by the following example:

Example 7.20

It is known that the conditional probabilities for the results for a new diagnostic screening test for a male disease are as follows:

	Man does have disease	Man does *not* have disease
Probability of positive result	$p(P|D) = 0.9$	$p(P|\overline{D}) = 0.2$
Probability of negative result	$p(\overline{P}|D) = 0.1$	$p(\overline{P}|\overline{D}) = 0.8$
Total probabilities	$p(P|D) + p(\overline{P}|D) = 1.0$	$p(P|\overline{D}) + p(\overline{P}|\overline{D}) = 1.0$

Assuming that 5 % (0.05) of men of age 60 do have the disease, calculate, for this age group:

(i)   the probability that a man selected at random would record a *true* positive result, (he must both have the disease and then record positive);

(ii) the probability that a man selected at random would record a *false* positive result, (he must be both free of the disease and then record positive);

(iii) the probability that a man selected at random would record a positive result (this will include the possibility of a true positive as well as a false positive);

(iv) the probability that a man selected at random, who tests positive, would actually have the disease.

See the following text for the calculation.

We perform the calculations for Example 7.20 by combining probabilities directly, but, for illustration, we also assume a random population of (say) 1000 men and work out the equivalent number of men who would fall into each category based on these probabilities.

(i) For a man, selected at random, to record a *true positive* result requires that the man must have the disease *AND* he must then test positive.
  - Probability that a man, selected at random, has the disease, $p(D) = 0.05$
    Hence, out of 1000 men, the number with the disease, $n(D) = 0.05 \times 1000 = 50$
  - Probability that a man, with the disease, will test positive, $p(P|D) = 0.9$
    Hence, the probability that a man, selected at random, will record a true positive result:

$$p(\text{true positive}) = p(P|D) \times p(D) \Rightarrow 0.9 \times 0.05 \Rightarrow 0.045$$

which is equivalent to a number of men (out of 1000):

$$n(\text{true positive}) = p(P|D) \times n(D) \Rightarrow 0.9 \times 50 \Rightarrow 45$$

(ii) For a man, selected at random, to record a *false positive* result requires that the man does *NOT* have the disease *AND* he must then test positive.
  - Probability that a man, selected at random, does not have the disease,

$$p(\overline{D}) = 1 - p(D) \Rightarrow 1 - 0.05 \Rightarrow 0.95$$

Hence, out of 1000 men, the number with no disease, $n(\overline{D}) = 0.95 \times 1000 \Rightarrow 950$
  - Probability that a man, free of the disease, will test positive, $p(P|\overline{D}) = 0.2$
    Hence, the probability that a man, selected at random, will record a false positive result:

$$p(\text{false positive}) = p(P|\overline{D}) \times p(\overline{D}) \Rightarrow 0.2 \times 0.95 \Rightarrow 0.19$$

which is equivalent to a number of men (out of 1000):

$$n(\text{false positive}) = p(P|\overline{D}) \times n(\overline{D}) \Rightarrow 0.2 \times 950 \Rightarrow 190$$

(iii) For a man, selected at random, to record a positive result (*either* true or false) requires that either a true positive *OR* a false positive is recorded:

$$p(\text{positive}) = p(\text{true positive}) + p(\text{false positive}) = 0.045 + 0.19 = 0.235$$

Hence the number of men (out of 1000) recording a positive result:

$$n(\text{positive}) = n(\text{true positive}) + n(\text{false positive}) \Rightarrow 45 + 190 \Rightarrow 235$$

(iv) For a man, *selected at random*, who then records positive, the probability that he does *in fact have the disease* will be given by:

$$\text{Probability of having disease} = \frac{p(\text{true positive})}{p(\text{true positive}) + p(\text{false positive})} \Rightarrow \frac{0.045}{0.235} \Rightarrow 0.191$$

We could also work out this probability by using the fact that, out of the 235 men who test positive, only 45 will actually have the disease, giving the probability:

$$\text{Probability of having disease} = \frac{45}{235} \Rightarrow 0.191$$

This example shows the care that must be taken with random screening tests. The possibility of false positives occurring in the majority of people who do not have the disease can make the probability of a false diagnosis quite high. In the above example, a man tested at random who gives a positive result has less than a 20 % chance of actually being ill.

Screening tests are normally conducted with people identified as already being 'at risk', and are then followed up by more accurate diagnostic techniques.

Q7.26

 Using the data from Example 7.20, calculate the probability that a man, selected at random, who then tests negative for the disease, could actually have the disease.

7.4.8 Probability Trees

In complex conditional probability problems, as in Example 7.20, it is often useful to describe the dependent probabilities using a probability tree (Figure 7.17).

Probability trees are particularly useful when the probabilities in the second stage of the analysis are conditional on the outcomes from the first stage of the analysis.

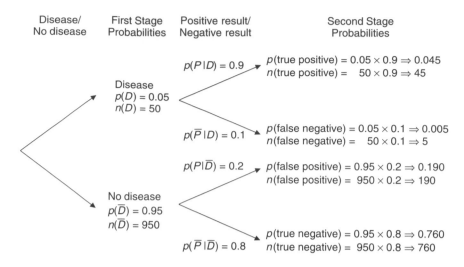

Figure 7.17   Probability tree for Example 7.20 based on 1000 men.

Q7.27

A bag initially contains three black sweets and four red sweets. In this problem two sweets are taken from the bag one at a time, and the first sweet is *not* returned to the bag.

Note that the probabilities for selection of the second sweet depend on what sweets are left after the first selection has been made.

Use a probability tree to work out the answers.

Calculate the probabilities of drawing:

  (i) a black sweet followed by another black sweet;
 (ii) a black sweet followed by a red sweet;
(iii) a red sweet followed by another red sweet;
 (iv) a red sweet followed by a black sweet;
  (v) any one of the above four combinations.

Q7.28

What is the probability that, out of five people, at least two of them have their birthdays in the same month? (Assume that every month $= 1/12$ of a year.)

Hint 1: it is easiest to calculate first the probability that no one has their birthday in the same month as anyone else.

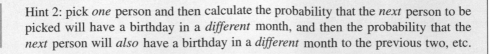

> Hint 2: pick *one* person and then calculate the probability that the *next* person to be picked will have a birthday in a *different* month, and then the probability that the *next* person will *also* have a birthday in a *different* month to the previous two, etc.

7.5 Factorials, Permutations and Combinations

7.5.1 Introduction

This unit develops the statistics for calculating the number of ways of obtaining particular arrangements of objects and events. These calculations have particular relevance in a range of probability calculations, from the likelihood of winning the National Lottery to the distribution of plants across a field.

7.5.2 Factorial

The **factorial**, $n!$, of a number, n, is the number multiplied by every integer between itself and one:

$$n! = n \times (n-1) \times (n-2) \times \cdots \times 3 \times 2 \times 1 \qquad [7.23]$$

Scientific calculators can calculate the factorial of a number directly – look for the $x!$ function.
   The factorial can be calculated in Excel (Appendix I) using the function FACT(x).
   It is useful to note some particular values:

$$1! = 1 \qquad\qquad\qquad\qquad\qquad\qquad\qquad\qquad\qquad [7.24]$$

$$n! = n \times (n-1)! \qquad\qquad\qquad\qquad\qquad\qquad\qquad [7.25]$$

$$0! = 1 \text{(This is not obvious, but should be remembered!)} \qquad [7.26]$$

Example 7.21

$1! = 1$

$2! = 2 \times 1! = 2 \times 1 = 2$

$3! = 3 \times 2! = 3 \times 2 = 6$

$4! = 4 \times 3! = 4 \times 3 \times 2 = 24$

etc.

$0! = 1$

$0!/3! = 1/6$

$20! = 20 \times 19!$

Division using factorials Using a simple numerical example first:

Example 7.22

$$\frac{8!}{5!} = \frac{8 \times 7 \times 6 \times \cancel{5} \times \cancel{4} \times \cancel{3} \times \cancel{2} \times 1}{\cancel{5} \times \cancel{4} \times \cancel{3} \times \cancel{2} \times 1} = \frac{8 \times 7 \times 6}{1} = 336$$

In Example 7.22, the 5! in the denominator cancels with all the terms from 5 to 1 in the numerator, leaving only $8 \times 7 \times 6$.

Hence, we can write a general formula for the division of factorials, where $n > m$:

$$\frac{n!}{m!} = n \times (n - 1) \times \cdots \times (m + 2) \times (m + 1) \qquad [7.27]$$

Example 7.23

Using equation [7.27]:

(i) $6!/4! = 6 \times 5 = 30$

(ii) $46!/41! = 46 \times (46 - 1) \times \cdots \times (41 + 2) \times (41 + 1) = 46 \times 45 \times 44 \times 43 \times 42$

Q7.29

Work out the values of each of the following *without* using a calculator (check the results afterwards using a calculator):

(i) $5!$

(ii) $(4 - 3)!$

(iii) $(3 - 3)!$

(iv) $3! - 3!$

(v) $\dfrac{7!}{5!}$

(vi) $\dfrac{101!}{99!}$

7.5.3 Permutations

A mathematical *permutation* is the *number of ways of arranging r items in order* when selected from n items.

Arranging *n* items in order The simplest problem is to find out how many ways it is possible to arrange *n* items into different sequences (or order).

Example 7.24

If we have three items, X, Y and Z, we can produce six possible arrangements which differ in the ordering of the items:

$$XYZ, XZY, YXZ, YZX, ZXY, ZYX$$

We can calculate the number of ways mathematically by considering how many 'choices' we have when filling the *first, second* and *third* places in order. Initially we have *three* choices for which letter to put in first place (e.g. Y), but we will then only have a choice of *two* letters left (e.g. X and Z) for the second place, and then only *one* letter for the third place:

We have a choice of three items for the first place (X, Y or Z)	We now have two choices left for the second place	We only have one 'choice' left for the final place	Total number of different ways
3	×2	×1	$= 3! = 6$

Thus the total number of possible arrangements was $3 \times 2 \times 1 = 3!$.

In general:

- -

Number of ways of arranging *n* items in order $= n!$ [7.28]

- -

Q7.30

In how many ways can the four nitrogen bases, A, T, G and C, be arranged in different orders?

Arranging *r* items in order when selected from *n* possible items We can illustrate the problem using the following example.

Example 7.25

As a specific example, consider selecting 3 $(= r)$ items, *in a specific order*, from 7 $(= n)$ items (e.g. TUVWXYZ). The number of ways can be calculated as follows:

We have a choice of $n = 7$ items for the first place	We now have $(7 - 1)$ choices left for the second place	We now have $(7 - 2)$ choices left for the third place	Total number of different ways
7	×6	×5	$= 7 \times 6 \times 5$

In this example the total number of ways of ordering 3 $(= r)$ from 7 $(= n)$ items is $7 \times 6 \times 5 = 210$.

We can rewrite the number as follows:

$$7 \times 6 \times 5 = \frac{7 \times 6 \times 5 \times 4 \times 3 \times 2 \times 1}{4 \times 3 \times 2 \times 1} = \frac{7!}{4!} = \frac{7!}{(7-3)!} = \frac{n!}{(n-r)!}$$

The general formula for arranging *r* items *in order* when selected from *n* possible items is given by the **permutation function**:

$$_nP_r = \frac{n!}{(n-r)!} \qquad [7.29]$$

This function can also be calculated directly in most scientific calculators. Enter the '*n*' value first, press the $_nP_r$ button, enter the '*r*' value, and press '='.

The $_nP_r$ function can be calculated in Excel (Appendix I) using the function PERMUT (n, r).

Example 7.26

Calculate the number of ways of selecting the first, second and third horses (order important) in a race with 10 horses:

$$_{10}P_3 = 10!/(10-3)! = 10 \times 9 \times 8 = 720$$

> **Q7.31**
>
> An athlete has eight different trophies, but only has room for four trophies in a display cabinet.
>
> How many different ways is it possible to display just four trophies out of eight, assuming that the display order is important?

7.5.4 Combinations

A mathematical *combination* is the number of different *ways of selecting* 'r' items from 'n' items when the order is *not* important.

Example 7.27

Calculate the number of ways of selecting 2 items from 4 possible items.

There are $12(= {}_4P_2)$ different ways of *ordering* 2 letters from the 4 letters W, X, Y and Z:

$$WX, WY, WZ, XW, XY, XZ, YW, YX, YZ, ZW, ZX, ZY$$

However, if the *order of selection* is *not* important, then there are *no differences* between each of the following 6 pairs of selections:

$$WX \text{ and } XW, WY \text{ and } YW, WZ \text{ and } ZW, XY \text{ and } YX, XZ \text{ and } ZX, YZ \text{ and } ZY$$

In this case there will then be only 6 *different* ways of selecting 2 items from 4 items with the order *not* important:

$$WX, WY, WZ, XY, XZ, ZY$$

The number of ways of *selecting* r items from n items when the order is not important is given by the function ${}_nC_r$.

We know that the number of ways that each *selection* of the r items can be ordered amongst themselves is $r!$. Hence, from the previous two statements, we see that the number of ways of first *selecting* and then *ordering* r items from n items, ${}_nP_r$, is given by the equation:

$$ {}_nP_r = {}_nC_r \times r! \qquad [7.30]$$

The number of combinations can therefore be calculated using:

$$_nC_r = \frac{_nP_r}{r!} \qquad\qquad [7.31]$$

which is the same as:

$$_nC_r = \frac{n!}{r! \times (n-r)!} \qquad\qquad [7.32]$$

This function can also be calculated directly in most scientific calculators. Enter the 'n' value first, press the $_nC_r$ button, enter the 'r' value, and press '='.

The $_nC_r$ function can be calculated in Excel (Appendix I) using the function COMBIN(n, r).

Example 7.27 can be solved using [7.32], to give the number of ways of selecting ($r =$) 2 items from ($n =$) 4 items (W, X, Y and Z) where the order is *not* important:

$$_4C_2 = \frac{4!}{2! \times (4-2)!} = \frac{24}{2 \times 2} = 6$$

Example 7.28

Calculate the number of ways of selecting the first, second and third horses in a race with 10 horses, where order is *not* important (see Example 7.26):

$$_{10}C_3 = {}_{10}P_3/3! = 720/6 = 120$$

Example 7.29

UK National Lottery (Lotto)

Calculate the probability of getting the jackpot in the UK National Lottery (Lotto), where 6 different numbers must be selected from 49 numbers.

The number of *arrangements* of 6 numbers selected from 49 numbers is:

$$_{49}P_6 = 1.0068 \times 10^{10}$$

However, many of the above arrangements will have selected the *same numbers but in a different order*. For each set of 6 numbers, there are 6! ways of rearranging them into a different order.

To win the lottery, the order of the six numbers is *not* important. The number of ways of selecting 6 numbers *in any order* from 49 numbers is:

$$_{49}C_6 = {}_{49}P_6/6! = 1.0068 \times 10^{10}/6! = 139\,838\,16 \approx 14\text{million}$$

The aim, when buying a lottery ticket, is to guess which one of the 14 million possible selections of the 6 numbers will occur.

Thus the probability of selecting the *correct* six numbers is 1 in 14 million!

Q7.32

A football manager has a squad of 20 players.

How many different teams of 11 players could be selected from the squad?

8

Distributions and Uncertainty

Overview

- 'How to do it' video answers for all 'Q' questions.
- Excel tutorials: statistical calculations, calculation of uncertainties.
- Excel files appropriate to selected 'Q' questions and Examples.

The concept of a continuous 'distribution' in statistics was introduced in 7.3.8, and is a method of describing how the values of a data variable may be spread over a given range. The distribution may record the variability of data that has been recorded in the past (frequency distribution) or predict the variability with which data values might be expected to be recorded in the future (probability distribution).

This chapter describes three distributions important in science – the normal, binomial and Poisson distributions. These have importance in modelling different systems in science:

- The normal distribution can be used to model a very wide range of systems where the distribution follows the familiar 'bell-shaped' curve. A particularly common application in science occurs in relation to experimental uncertainty.
- The binomial distribution applies to systems where, at the fundamental level, the system can be in *one of two* possible states – hence the prefix 'bi'. A classic example of binomial statistics might calculate the probability of getting eight heads when a coin is tossed 10 times.
- The Poisson distribution is a special case of the binomial distribution, which occurs when the individual probability of a given event is low. This distribution can be used to estimate the probabilities for the occurrences of such events, e.g. the probability that a particular number of cases of a randomly occurring disease may appear in a specific locality.

The most important parameters used to describe a distribution – mean, standard deviation, variance – have already been introduced in 7.2.4 and 7.2.6, and are used to describe the *location* and *spread* of the distribution.

Other parameters that are used to describe the characteristic shape of distributions are given on the Website.

Essential Mathematics and Statistics for Science 2nd Edition   Graham Currell and Antony Dowman
Copyright © 2009 John Wiley & Sons, Ltd

Different types of scientific mechanisms may produce experimental data with different distributions. If it is possible to confirm that the data falls into one or other type of distribution, then it is possible to deduce that one mechanism is more likely to be operating than a possible alternative. In this respect, 'goodness of fit' (14.2) is a useful hypothesis test that can be used to assess whether an experimental distribution might be different from a particular theoretical model.

The impact of experimental uncertainty on scientific measurements has been introduced in 1.2. The normal distribution has particular value in dealing with experimental uncertainty, as it closely models the variability in the results of many types of experimental measurement. In particular, the random uncertainty can be quantified by defining a *confidence interval* at X%, which has an X% probability of including the true value being measured.

For example, if an experimental result is presented as:

$$4.67 \pm 0.05 \ (95\% \ \text{CI})$$

then it can be stated with 95 % probability that the true value of the variable being measured lies between 4.62 ($= 4.67 - 0.05$) and 4.72 ($= 4.67 + 0.05$).

The ways in which the uncertainties in separate variables combine in a calculation of overall uncertainty is discussed in 8.3.3.

8.1 Normal Distribution

8.1.1 Introduction

The bell-shaped normal distribution is the most widely used distribution in science. Of particular importance is the fact that many experimental variations can be described effectively by using the normal distribution.

8.1.2 Experimental uncertainty

There is always some uncertainty when making any experimental measurement. If we are trying to measure some experimental property whose true value is μ, then we are more likely to record a value close to μ, and less likely to record a value a long way away from μ.

The probability density, $pd(x)$ (7.3.8), of recording a particular value, x, is largest when x is close to μ, but then the probability falls off as a 'bell-shaped' curve on either side of μ.

Example 8.1

Figure 8.1 shows the probabilities of getting an experimental result, x, in specific ranges, when measuring a **true** blood–alcohol level of 80 mg per 100 mL, i.e. 80 mg of alcohol in each 100 mL of blood. The *standard deviation* uncertainty in the measurement process being used is 2.0 mg per 100 mL.

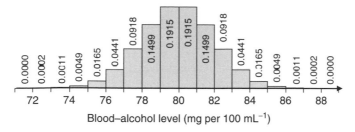

Figure 8.1 Histogram of blood–alcohol measurements.

For example, the histogram shows that the probability of recording a value between 81 and 82 is 0.1499 (or about 15 %).

The mathematical name for this 'bell-shaped' distribution is the **normal distribution**.

Most *random* experimental errors can be considered to have a frequency spread of results that follow a normal distribution.

Q8.1

Figure 8.1 shows the probabilities of a single experimental result, with a standard deviation of 2.0, falling within unit ranges about a mean value of 80.

Calculate, *using the data* in Figure 8.1, the probabilities that a single experimental result will fall within the following ranges (in mg per 100 mL):

(i) 80–81
(ii) 81–83
(iii) >80
(iv) 78–82 (i.e. 80 ± 1 standard deviation)
(v) 76–84 (i.e. 80 ± 2 standard deviations)
(vi) 74–86 (i.e. 80 ± 3 standard deviations)
(vii) 72–88 (i.e. 80 ± 4 standard deviations)

8.1.3 Normal distribution

The normal distribution (also called the Gaussian distribution) is defined by the probability density, $pd(x)$, as a function of the variable, x (with mean value, μ, and standard deviation, σ).

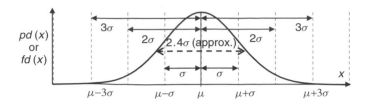

Figure 8.2   Normal distribution with mean, μ, and standard deviation, σ.

The normal curve in Figure 8.2 gives the probability density, $pd(x)$, or frequency density, $fd(x)$, of recording a particular value, x, when the values are distributed with a standard deviation, σ, about a mean, μ.

Example 8.2

Make a careful note of the 'width' of the curve in Figure 8.2 compared with the size of the standard deviation, σ. That is:

- There are *about* three standard deviations, 3σ, between the centre of the distribution and the extreme 'tail' on each side.
- The distribution *width at half its maximum height* is approximately equal to $2.4 \times \sigma$.

There are very many different normal curves, each defined by a different *mean* and *standard deviation*. However, two important facts are that:

- they all have the same overall 'bell' *shape*; and
- they all have the same *area* (in the case of the probability distribution, this area is 1.0).

Figure 8.3 shows normal distributions with means 12 and 18 and standard deviations 2 and 4 respectively. Note that they have the same probability area ($= 1.0$), with one curve having twice the height of the other but half the width.

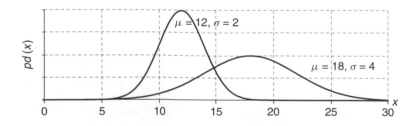

Figure 8.3   Normal distributions.

Q8.2

(i) Estimate, by eye, the *mean* of the normal distribution given in Figure 8.4.

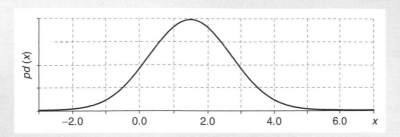

Figure 8.4

(ii) Estimate the *standard deviation* by using the fact there are approximately three standard deviations between the centre and each extreme tail.

(iii) Estimate the *standard deviation* by using the fact that the width of a normal distribution at half its maximum height is approximately equal to $2.4 \times \sigma$, where σ is the standard deviation. (On the graph in the figure, estimate the width of the distribution at half of the maximum height of the curve, and then divide the value by 2.4.)

8.1.4   Common probability areas

Figure 8.5 shows the probabilities (areas) of recording values in specific regions defined by the 'numbers of standard deviations' from the mean value.

The probability of recording a value between x_1 and x_2 can be written as $p(x_1 < x < x_2)$.

The *probability*, $p(x_1 < x < x_2)$, is equal to the *area* under the standard normal distribution between values x_1 and x_2.

The *total probability* (*area*) for *all* values of x, $p(-\infty < x < +\infty) = 1.0$, and the probability area for each 'half' of the distribution, $p(-\infty < x < 0) = p(0 < x < +\infty) = 0.5$.

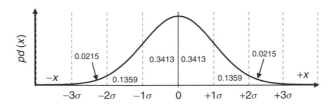

Figure 8.5   Some probability areas for a normal distribution.

The normal distribution is symmetrical – the areas on the negative side of the distribution are the same as the equivalent areas on the positive side.

Calculations involving the probability areas of a normal distribution can be performed in Excel or by reading probability areas directly from the table in Appendix II – see Appendix I and the Website.

Example 8.3

Using the probability areas in Figure 8.5, we can calculate the probabilities of recording values of x within specific ranges (it is often necessary to add or subtract areas to get the desired final area):

(i) Probability of recording a value within *one* standard deviation ($\pm 1\sigma$) of the mean:

$$p(-1\sigma < x < +1\sigma) = 2 \times 0.3413 = 0.6826 \approx 68.3\,\%$$

(ii) Probability of recording a value within *two* standard deviations ($\pm 2\sigma$) of the mean:

$$p(-2\sigma < x < +2\sigma) = 2 \times (0.3413 + 0.1359) = 0.9544 \approx 95.4\,\%$$

(iii) Probability of recording a value within *three* standard deviations ($\pm 3\sigma$) of the mean:

$$p(-2\sigma < x < +2\sigma) = 2 \times (0.3413 + 0.1359 + 0.0215) = 0.9974 \approx 99.7\,\%$$

Incorporating the results from Example 8.3, we can list in Table 8.1 some common probabilities associated with *all* normal distributions.

Table 8.1.   Useful probability areas for the normal distribution.

68.3%	of all data points lie within	$\pm 1.00 \times \sigma$ of the mean value
90%	of all data points lie within	$\pm 1.64 \times \sigma$ of the mean value
95%	of all data points lie within	$\pm 1.96 \times \sigma$ of the mean value
95.4%	of all data points lie within	$\pm 2.00 \times \sigma$ of the mean value
99%	of all data points lie within	$\pm 2.58 \times \sigma$ of the mean value
99.7%	of all data points lie within	$\pm 3.00 \times \sigma$ of the mean value

---

Q8.3

It is known that a specific measurement process produces results which are normally distributed with a standard deviation of 4.

If 100 000 replicate measurements could be made of a true value equal to 50, use the information in Table 8.1 to estimate the number of results that would record a value:

(i) between 46 and 54
(ii) between 42.16 and 57.84
(iii) outside the range 38 to 62
(iv) greater than 54 (hint: use your result from (i))

8.2 Uncertainties in Measurement

8.2.1 Introduction

We have seen in 1.2 that a typical experimental measurement will have a random uncertainty, which may be due to either:

- a *variation in the measurement procedure* itself (e.g. uncertainty in the response of a pH electrode); or
- a *natural variation in the subjects* being measured (e.g. different plants of the same crop grow at different rates).

The problems of experimental uncertainty are fundamental to all experimental science, and a more extensive discussion of experimental variation is given in 11.1.2.

8.2.2 Experimental variation and true value

In a typical experimental measurement, we aim to try to identify the *unknown value, μ*, of some scientific parameter (e.g. pH of a particular solution, or the average growth rate of crop plants).

The **true value** (1.2) means something slightly different in each of the two types of experimental uncertainty:

- Where the uncertainty is due entirely to the measurement process itself, the true value is the *single* actual value, μ, of the subject that is being measured (e.g. the true pH of the solution).
- Where the major uncertainty lies in the subjects that are being measured, the true value is the mean value, μ, of *all* the possible measurements (population) that could be made (e.g. the mean growth rate of all the plants of a given crop).

Fortunately, we can apply the *same statistics* to both of the above situations. We see that the mean value, $\bar{x}$, of a *sample* of n experimental measurements is used as a 'best estimate' for true mean value, μ, that would be obtained if a full *population* (see 7.2.2) of measurements could be made.

A population of measurements for *measurement* uncertainty would be an infinite number of replicate (repeated) measurements, and a population of measurements for *subject* uncertainty would include every possible measurement (e.g. every plant in the crop).

Q8.4

Consider the two experimental investigations:

A: a metallurgist makes four repeated (replicate) analyses for the concentration of cadmium in a piece of iron.
B: a biologist measures the lengths of 30 stems from a crop of soya bean.

Both of these scientists will record a mean value, $\bar{x}$, and a standard deviation, s, for their experimental results.

(i) Explain, in both cases, the probable sources of the variation, s, in the experimental results.
(ii) What will be the relevance of the value, $\bar{x}$, to the metallurgist in A?
(iii) What will be the relevance of the value, $\bar{x}$, to the biologist in B?

The effect of experimental variation can be described by using the probability density, $pd(x)$, of recording a particular experimental value, x.

Figure 8.6 shows that the *true value* being measured is μ, but experimental uncertainty means that the *probability* of getting an experimental *value*, x, is often given by a *bell-shaped* normal distribution.

The distribution shows that it is more likely that the experimental result, x, will be close to μ, but there is also a decreasing probability that a result, x, might be recorded a long way from μ.

From 8.1.7 we can see that, for *individual* experimental results, x:

90 % of x would fall in the range $\mu - 1.64 \times \sigma$ to $\mu + 1.64 \times \sigma$

95 % of x would fall in the range $\mu - 1.96 \times \sigma$ to $\mu + 1.96 \times \sigma$ [8.1]

99 % of x would fall in the range $\mu - 2.58 \times \sigma$ to $\mu + 2.58 \times \sigma$

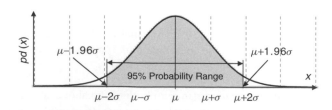

Figure 8.6 Distribution of individual results, x, for an experimental uncertainty which has a normal distribution with a mean, μ, and standard deviation, σ.

It is worth remembering the numbers 1.64, 1.96 and 2.58 – they occur frequently.

Q8.5

In a particular experiment to measure the concentration of lead in drinking water, the experimental procedure has a standard deviation uncertainty of 2.0 ppb. A number of analysts each make a *single* experimental measurement.

If the true value of the concentration is 23 ppb, estimate the range of values within which 95 % of the analysts' results are likely to fall.

8.2.3 Replicate measurements, sample means and standard error

The uncertainty in a *single* experimental measurement is given directly by the standard deviation, σ, of the experimental variation – see 8.2.2. However, it is good scientific practice to take several repeat (or *replicate*) measurements, n, and then take the average (mean) of that sample of n measurements.

When we take an increasing number of replicate measurements, the mean, $\overline{x}$, of our replicates will tend to become closer to the 'true' value, μ. This is a consequence of a statistical theory called the 'central limit theorem'.

Another important consequence of the central limit theorem is that the distribution of *sample mean* values will tend towards a *normal* distribution, even if the distribution of the individual measurements is *not normal* (e.g. a Poisson distribution). By taking the mean values of data samples, we are more likely to ensure that our experimental uncertainties do follow normal distributions – this is important for the validity of much of the statistics used to analyse such data. In practice, a sample size of at least 30 data values is usually sufficient.

The reduced uncertainty (or *improved* precision) when taking the *mean* of n *replicate measurements* is given by:

$$\text{Standard error of the mean, } SE = \frac{\sigma}{\sqrt{n}} \qquad [8.2]$$

Note that the equivalent statistic when using the sample standard deviation, s, is called the standard uncertainty – see [8.4].

According to the central limit theorem, the distribution of the possible mean values, $\overline{x}$, of n experimental measurements follows a normal distribution, as in Figure 8.7, with a mean (of mean values) $= \mu$ and a standard deviation of $SE = \sigma/\sqrt{n}$.

Figure 8.7 gives the probability density, $pd(\overline{x})$, for recording mean values, $\overline{x}$, of different sample sizes, n. The sample size $n = 1$ represents a single measurement as in Figure 8.6.

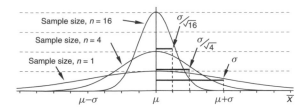

Figure 8.7   Distribution of mean values, $\bar{x}$, for samples of different sizes.

As the sample size, n, increases:

- the uncertainty in the mean value gets smaller;
- the probability curve gets narrower; and
- the experimental result, $\bar{x}$, is likely to be closer to the 'true' value, μ.

Comparing with equations [8.1], we can see that, for the *mean values*, $\bar{x}$, of n experimental measurements (note the effect of the number of replicate measurements, n):

90 % of $\bar{x}$ - values would fall in the range $\mu - \left(1.64 \times \dfrac{\sigma}{\sqrt{n}}\right)$ to $\mu + \left(1.64 \times \dfrac{\sigma}{\sqrt{n}}\right)$

95 % of $\bar{x}$ - values would fall in the range $\mu - \left(1.96 \times \dfrac{\sigma}{\sqrt{n}}\right)$ to $\mu + \left(1.96 \times \dfrac{\sigma}{\sqrt{n}}\right)$

$$[8.3]$$

99 % of $\bar{x}$ - values would fall in the range $\mu - \left(2.58 \times \dfrac{\sigma}{\sqrt{n}}\right)$ to $\mu + \left(2.58 \times \dfrac{\sigma}{\sqrt{n}}\right)$

Q8.6

In the same experimental procedure as in Q8.5 the standard deviation uncertainty of measurement is 2.0 ppb. In this case, each analyst makes four measurements and then calculates the mean value, $\bar{x}$, of their four measurements:

(i) Calculate the standard error of the mean values, $\bar{x}$.
(ii) If the true value of the concentration is 23 ppb, estimate the range of values within which 95 % of the experimental means, $\bar{x}$, are likely to fall.

8.2.4   Experimental perspective and confidence intervals

The discussion of the normal distribution above assumes that the true (or population) mean value, μ, is known and that the standard deviation of the distribution, σ, is also known.

However, in most experimental measurements it is the unknown value, μ, that is being measured and very often the inherent standard deviation, σ, of data values is also unknown.

The typical experimental measurement with n replicate measurements will report a sample mean value, $\bar{x}$, and a sample standard deviation, s:

- $\bar{x}$ is the 'best estimate' for the true mean value, μ;
- s is the 'best estimate' for the true standard deviation, σ.

The best estimate for the inherent uncertainty in a *single* experimental measurement is given by the sample standard deviation, s.

The resultant uncertainty in the final mean value, $\bar{x}$, is reduced by having taken n replicate measurement. On the basis of the central limit theorem (8.2.3) and equation [8.2] which describes how the uncertainty in the mean is reduced as the square root of the number of data values, we can then define:

$$\text{Standard uncertainty in experimental value, } u_x = \frac{s}{\sqrt{n}} \qquad [8.4]$$

It would now be possible to report the results of the experiment, giving the observed mean value, $\bar{x}$, and quoting the standard uncertainty, $s/\sqrt{n}$.

For example, if nine replicate measurements gave a mean value of 8.7 and a sample standard deviation of 0.6, the results *could* be quoted as a best estimate for the true value of 8.7 with a standard uncertainty of $0.6/\sqrt{9} = 0.2$.

However, it is even more useful to be able to quote the experimental results in a statement which gives a calculated range of possible values:

With 95 % confidence the true value being measured is within the range 8.7 ± 0.46, i.e. between $8.7 - 0.46$ (≈ 8.2) and $8.70 + 0.47$ (≈ 9.2).

The range quoted above is called the '95 % confidence interval' and is a standard method of communicating these results.

The **confidence interval**, $CI_{X\%}$, is defined as the *total range* within which we are $X\%$ confident in stating that the true value, μ, lies:

$$CI_{X\%} = \bar{x} \pm \left(t_{2,\alpha,df} \times \frac{s}{\sqrt{n}} \right) \qquad [8.5]$$

The confidence interval is centred on the sample mean, $\bar{x}$, and extends a distance on either side, called the **confidence deviation**:

$$Cd_{X\%} = t_{2,\alpha,df} \times \frac{s}{\sqrt{n}} \qquad [8.6]$$

The new factor in equations [8.5] and [8.6] is the '2-tailed' t-value, $t_{2,\alpha,df}$ (see also 10.1.3), where

α is the *significance level* (see also 9.4.4) with: $\alpha = 1 - \frac{X\%}{100}$
(typically $\alpha = 0.05$ when $X\% = 95\%$) and

df is the *degrees of freedom* for the calculation.

Degrees of freedom is a statistical concept dependent on the number of separate bits of information available for the calculation. For the confidence interval calculation the number of degrees of freedom is given by $df = n - 1$, where n is the number of data values in the sample.

A selection of '2-tailed' t-values, $t_{2,\alpha,df}$, for 95 % confidence ($\alpha = 0.05$) is given in Table 8.2 for a number of sample sizes. A more extensive table of t-values is given in Appendix III.

Table 8.2.

Sample size, n	2	3	4	5	6	8	11	21	∞
Degrees of freedom, df	1	2	3	4	5	7	10	20	∞
$t_{2,0.05,df}$	13.7	4.30	3.18	2.78	2.57	2.36	2.23	2.09	1.96

The variation of the t-value is due to the fact that the calculation is based on the assumption that the sample standard deviation, s, is the best estimate for the true standard deviation, σ, of the data values. This has two consequences that can be seen in Table 8.2:

- For large sample sizes (n large), s will become a good approximation for σ, and the t-value approaches 1.96, which is the numerical value in equations [8.3] for the 95 % range of data values calculated using σ.
- For small sample sizes (n small), there will be increasing uncertainty that s is a good approximation for σ, and this increasing uncertainty must be reflected in a wider confidence interval, which in turn requires a larger value for $t_{2,0.05,df}$.

The confidence interval is equivalent to the range:

$$\bar{x} - Cd_{X\%} \text{ to } \bar{x} + Cd_{X\%}$$

or:

$$\bar{x} - \left(t_{2,\alpha,df} \times \frac{s}{\sqrt{n}}\right) \text{ to } \bar{x} + \left(t_{2,\alpha,df} \times \frac{s}{\sqrt{n}}\right) \qquad [8.7]$$

It is interesting to note that the statistical calculations for the confidence interval based on t-values were first developed by William Gosset to analyse small statistical samples arising from measurements being made in the brewing of Guinness in Dublin in 1908. Due to industrial secrecy he published his work under the pseudonym 'Student'.

Note that the Excel function CONFIDENCE (Appendix I) should **not be used** for *sample standard deviations*. The function assumes that the population standard deviation, σ, is known, which is not normally true in an experimental context.

Example 8.4

A set of $n = 6$ replicate measurements is made, giving experimental results: 25, 28, 23, 20, 25, 24.

Calculate the range of values within which we can be 95 % confident that the unknown true value, μ, lies.

The mean of the six results, $\bar{x} = 24.17$.

The calculated sample standard deviation of the six replicates, $s = 2.64$ (this is our best estimate for the unknown population standard deviation, σ).

From Table 8.2, a sample size $n = 6$ gives a value of $t_{2,0.05,5} = 2.57$:

$$CI_{95\%} = \bar{x} \pm \left(t_{2,0.05,5} \times \frac{s}{\sqrt{n}}\right) \Rightarrow 24.17 \pm \left(2.57 \times \frac{2.64}{\sqrt{6}}\right) \Rightarrow 24.17 \pm 2.77$$

We can then say with 95 % confidence that, on the basis of six replicate measurements, which have a mean of 24.17 and a sample standard deviation of 2.64, the true value being measured lies between 21.4 and 27.0 (rounded to 1 dp).

Q8.7

A student develops a new quick technique to measure the concentration of cadmium (Cd) in river water, and uses the technique to record the values given below (in ppb) for an unknown sample. In each of the cases below the student wishes to report the true concentration for the unknown sample, with a 95 % probability of being correct.

(i) The student makes three replicate measurements: 18.2, 20.2, 19.6. What can the student say about the true value?

(ii) The student makes six replicate measurements: 18.2, 20.2, 19.6, 17.6, 18.5, 19.5. What can the student say about the true value?

(iii) The student records only one value: 18.2. What can the student say about the unknown true value?

Example 8.5

Two analysts are both measuring the concentration of the same solution and report their individual findings for their 95 % confidence intervals as 18.6 ± 1.2 and 14.9 ± 2.6, respectively.

Is it possible for the true value to fall within *both* confidence Intervals?

The two confidence intervals overlap between the values 17.4 and 17.5, so that a true value within that range would also be in both confidence intervals.

8.3   Presenting Uncertainty

8.3.1   Introduction

The management of uncertainty in experimental measurements is a recurrent theme throughout this book. In mathematical terms, the aim of a typical measurement is to determine the true value, μ (population mean), of a given variable, for which the scientist makes n replicate measurements of the variable, x, to give an experimental mean value, $\bar{x}$, with a sample standard deviation, s.

Within this unit, we start by introducing some of the principal terms that are used by scientists to *present* this experimental uncertainty. We then develop the methods that can be used to *combine* the separate uncertainties of individual measurements into a single uncertainty in a final experimental result.

8.3.2   Terminology

It is useful to review some of the terms that we use for describing the uncertainty in a particular experimental measurement.

Standard uncertainty, u_x, in the mean value, $\bar{x}$, of n measurements of a variable, x, is defined by the International Organization for Standardization (ISO) *Guide to the Expression of Uncertainty in Measurement* as the standard deviation, s, of the measurements divided by the square root of n:

$$u_x = \frac{s}{\sqrt{n}} \tag{8.8}$$

The standard uncertainty (see also 8.2.4) is the best estimate of the standard error of the mean, *SE* (8.2.3), of the population of measurements from which the sample of experimental measurements was drawn.

We define a new term, **relative uncertainty**, Ru_x, in the mean value, $\bar{x}$, as the percentage ratio of the standard uncertainty to the mean value. Relative uncertainty is the percentage uncertainty in a *particular* experimental *result*:

$$Ru_x = 100 \times \frac{u_x}{\bar{x}} \text{ and } u_x = \frac{\bar{x} \times Ru_x}{100} \qquad [8.9]$$

We have defined, in 8.2.4, a new term, **confidence deviation (Cd)**, which is the range *in either direction from the mean value*, $\bar{x}$, within which the true value, μ, can be expected to lie, with a confidence of $X\%$:

$$Cd_{X\%} = t_{2,\alpha,df} \times \frac{s}{\sqrt{n}} \qquad [8.10]$$

Confidence interval (CI) is the range of values (8.2.4) within which the true value, μ, can be expected to lie, with a confidence of $X\%$:

$$CI_{X\%} = \bar{x} \pm Cd_{X\%}$$

$$CI_{X\%} = \bar{x} \pm \left(t_{2,\alpha,df} \times \frac{s}{\sqrt{n}} \right) \qquad [8.11]$$

It is also useful to note two terms that are used to describe the *precision* of an experimental *process* rather than the uncertainty in a particular measurement: **relative standard deviation (RSD)** and **coefficient of variation (CV)** both give the ratio of the population standard deviation, σ, to the mean value, i.e. the *fractional* standard deviation. The coefficient of variation is the 'percentage' equivalent of relative standard deviation:

$$RSD = \frac{\sigma}{\bar{x}}$$

$$CV = 100 \times \frac{\sigma}{\bar{x}}\% \qquad [8.12]$$

When presenting results, the experimental method used should be made explicit, and sufficient information (e.g. sample size, n) should be given to enable the readers to interpret the uncertainty in any of the formats given. The final result should be given with an appropriate number of significant figures (2.1.7).

Example 8.6

Nine experimental measurements give a mean value of 56.26 with a sample standard deviation of 5.7.

(i) Calculate values for the standard uncertainty, u_x, and relative uncertainty, Ru_x.
(ii) Express the final result using a 95 % confidence interval.

Here:

(i)
$$u_x = \frac{5.7}{\sqrt{9}} \Rightarrow 1.9$$

$$Ru_x = 100 \times \frac{1.9}{56.26} \Rightarrow 3.38\,\%$$

(ii) Degrees of freedom, $df = n - 1 \Rightarrow 9 - 1 \Rightarrow 8$
t-value for two-tailed, 95 %, 8 degrees of freedom $= 2.31$:

$$CI_{95\,\%} = 56.26 \pm 2.31 \times 1.9 \Rightarrow 56.26 \pm 4.389$$

which would give a final result, with rounding, expressed as:

$$\mu = 56.3 \pm 4.4 \ (95\,\%CI)$$

Q8.8

A particular measurement procedure is quoted as having a coefficient of variation [8.12] of 5 %. Calculate the standard error of the mean [8.2] that would be expected for the mean of eight replicate measurements of a true value of 4.5 ppm.

8.3.3 Combining uncertainties

There are many occasions in science when a final result, x, will be given by the combination of two or more values, e.g. a, b. The initial uncertainties in a, b can be expressed as:

- standard uncertainties, u_a and u_b; and/or
- relative uncertainties:

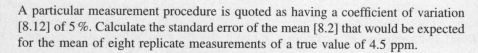

$$Ru_a = 100 \times \frac{u_a}{a} \text{ and } Ru_b = 100 \times \frac{u_b}{b}$$

The ways in which we combine the uncertainties depend on the way in which the values a and b are combined.

For *addition* and *subtraction* equations we combine the *squares* of *standard* uncertainties, but for *multiplication* and *division* we combine the *squares* of *relative* uncertainties:

If $x = k \times a$ (k is a constant) then use: $u_x = k \times u_a$ or $Ru_x = Ru_a$ [8.13]

If $x = a + b$ or $x = a - b$, then use: $u_x = \sqrt{u_a^2 + u_b^2}$ [8.14]

If $x = a \times b$ or $x = \dfrac{a}{b}$, then use: $Ru_x = \sqrt{Ru_a^2 + Ru_b^2}$ [8.15]

If $x = a^n$ (n is a constant) then use: $Ru_x = n \times Ru_a$ [8.16]

In performing calculations, it is often necessary to change between standard uncertainty and relative uncertainty using equations [8.9].

Example 8.7

The measurements of the two sides of a rectangle record the values:

$$a = 8.4 \text{ mm with a standard deviation uncertainty, } u_a = 0.5 \text{ mm}$$
$$b = 6.7 \text{ mm with a standard deviation uncertainty, } u_b = 0.5 \text{ mm}$$

Calculate the following values, *together with both the standard and relative uncertainties* in each value:

(i) Sum of the two lengths: $t = a + b$
(ii) Difference in the lengths: $d = a - b$
(iii) Perimeter of the rectangle: $p = 2(a + b)$
(iv) Ratio of the lengths of the sides: $r = a/b$
(v) Area of the rectangle: $A = a \times b$

Answers are given in the following text.

We can answer Example 8.7 by using equations [8.13] to [8.16]:

(i) The addition of the two lengths gives $t = 8.4 + 6.7 = 15.1$ mm.
As this is an 'addition', we must calculate the uncertainty, u_t, by using [8.14]:

$$u_t = \sqrt{0.5^2 + 0.5^2} = 0.707 \text{ mm}$$

The relative uncertainty is: $Ru_t = 100 \times \frac{u_t}{t} \Rightarrow 100 \times \frac{0.707}{15.1} \Rightarrow 4.68\,\%$

(ii) The difference between the two lengths gives $d = 8.4 - 6.7 = 1.7$ mm.
As this is a 'subtraction', we still calculate the uncertainty, u_t, by using [8.14]:

$$u_t = \sqrt{0.5^2 + 0.5^2} = 0.707 \text{ mm}$$

The relative uncertainty is: $Ru_t = 100 \times \frac{u_t}{t} \Rightarrow 100 \times \frac{0.707}{1.7} \Rightarrow 41.6\,\%$
In (i) and (ii), notice that, although the absolute error, u, is the same, the relative error, Ru_t, becomes much larger in (ii) because the actual value is much less (1.7 instead of 15.1).

(iii) To calculate the perimeter, $p = 2(a + b)$, we can use $t = a + b$ from (i), where $t = 15.1$ mm and $u_t = 0.707$, and calculate $p = 2t \Rightarrow 30.2$ mm.
As we are multiplying by a constant, we use [8.13]:

$$u_p = 2 \times u_t = 2 \times 0.707 = 1.414 \text{ mm}$$

The relative uncertainty is: $Ru_p = 100 \times \frac{u_p}{p} \Rightarrow 100 \times \frac{1.414}{30.2} \Rightarrow 4.68\,\%$
Note that the relative uncertainty has stayed the same (4.7 %) when simply multiplying t by the constant.

(iv) The ratio of the sides gives: $r = \frac{a}{b} \Rightarrow \frac{8.4}{6.7} \Rightarrow 1.254$ (pure number with no units)
As this is a 'division', we must use [8.9] to calculate the relative uncertainties of a and b:

$$Ru_a = 100 \times \frac{u_a}{a} \Rightarrow 100 \times \frac{0.5}{8.4} \Rightarrow 5.952\,\%$$

$$Ru_b = 100 \times \frac{u_b}{b} \Rightarrow 100 \times \frac{0.5}{6.7} \Rightarrow 7.463\,\%$$

Using [8.15], the relative uncertainty is:

$$Ru_r = \sqrt{Ru_a^2 + Ru_b^2} \Rightarrow \sqrt{5.942^2 + 7.463^2} \Rightarrow 9.55\,\%$$

Using [8.9] to calculate the absolute uncertainties:

$$u_r = \frac{r \times Ru_r}{100} \Rightarrow \frac{1.254 \times 9.55}{100} \Rightarrow 0.120 \text{ mm}$$

(v) The area of the rectangle gives $A = a \times b = 8.4 \times 6.7 = 56.28$ mm^2.
As this is a 'multiplication', and using [8.15] will give us the same relative uncertainty as for the 'division' in (iv), the relative uncertainty is:

$$Ru_A = \sqrt{Ru_a^2 + Ru_b^2} \Rightarrow \sqrt{5.942^2 + 7.463^2} \Rightarrow 9.55\%$$

Using [8.9] to calculate the absolute uncertainties:

$$u_A = \frac{A \times Ru_A}{100} \Rightarrow \frac{56.28 \times 9.55}{100} \Rightarrow 5.37 \text{ mm}^2$$

Q8.9

Calculate (with the combined uncertainty) the density, ρ, of a piece of compact bone which has been measured to have mass $m = 4.3$ g with $u_m = 0.3$ g and volume $V = 2.3$ cm^3 with $u_V = 0.2$ cm^3. Use the equation:

$$\rho = \frac{m}{V}$$

Use the table below to set out the results.

Variable	Value, x	u_x	Ru_x
Mass, m(g)			
Volume, V(cm^3)			
Density, ρ (g cm^{-3})			

Q8.10

Using the same data as in Example 8.7, calculate the value and uncertainty for the diagonal of the rectangle, $h = \sqrt{a^2 + b^2}$ (using Pythagoras), given that:

$a = 8.4$ mm with a standard deviation uncertainty, $u_a = 0.5$ mm

$b = 6.7$ mm with a standard deviation uncertainty, $u_b = 0.5$ mm.

8.4   Binomial and Poisson Distributions

8.4.1   Introduction

The binomial distribution applies to a situation where, for each particular *statistical trial*, there are only *two possible results* (e.g. Y or N, 1 or 0, X or $\overline{X}$ – hence the prefix 'bi'). For example:

- A room might contain 10 randomly selected people, each of whom will be either right-handed or left-handed – binomial statistics can calculate the probability that exactly three people will be left-handed.
- In an experiment to toss a coin 100 times, each toss of the coin is a *trial* – the binomial distribution describes the probabilities of obtained specific numbers of heads in the overall experiment (see Figure 9.1).

The Poisson distribution is a special case of the binomial distribution where the probability, p, of one outcome of each particular statistical trial is very low. For example, it could be appropriate to use the Poisson distribution to calculate the probability of occurrence of random genetic mutations.

With both the binomial and Poisson distribution, it is assumed that the result of each trial is *independent* of the results of any other trial, and that the subject of each trial is randomly selected.

8.4.2   Binomial distribution

We will use the following example as an introduction to the distribution.

Example 8.8

The parents of five offspring both carry the gene for albinism, which gives each offspring a 1 in 4 probability ($p = 0.25$) of being an albino.

We wish to calculate the probability, $p(r)$, that, within the group of $n = 5$ offspring, there will be exactly $r = 2$ albinos.

The worked answer is given in the following text.

In a typical binomial problem with a total of n *trials*, each trial could have an *outcome*, Y or N.

The term **trial** is often used in statistics, and, in Example 8.8, refers to a test as to whether a particular offspring is an albino (Y) or not (N).

If r trials have an outcome Y, then $(n-r)$ trials will have an outcome N.

If p is the probability that *each* trial will result in Y, then the probability, $p(r)$, that there will be exactly r trials with outcome Y is given by:

$$p(r) = {}_nC_r \times p^r \times (1-p)^{(n-r)} \qquad [8.17]$$

where the combination function, $_nC_r$, (7.5.4), is often also called the *binomial coefficient*.

The Excel function BINOMDIST(r, n, p,FALSE) can be used to calculate values for the probabilities of a binomial distribution. Entering 'TRUE', instead of 'FALSE', in the last argument value would give the *cumulative* probability of $p(r)$.

The derivation of equation [8.17] is given on the Website.

In Example 8.8, we can substitute $r = 2$, $n = 5$ and $p = 0.25$ into equation [8.17], giving the probability that there will be exactly ($r =$) 2 albinos in the five offspring:

$$p(2) = {_5C_2} \times 0.25^2 \times (1 - 0.25)^{(5-2)} \Rightarrow 10 \times 0.25^2 \times 0.75^3 \Rightarrow 0.2637$$

Similarly the probability of finding exactly ($r =$) 4 albinos would be given by:

$$p(4) = {_5C_4} \times 0.25^4 \times (1 - 0.25)^{(5-4)} \Rightarrow 10 \times 0.25^4 \times 0.75 \Rightarrow 0.0146$$

Q8.11

Repeat the calculation in Example 8.8 for each of the probabilities for 0, 1, 2, 3, 4 and 5 albinos.

Calculate the total of all of these probabilities.

Example 8.9

It is known that there are 1500 plants randomly distributed over an area of 1000 m^2.

(i) Use the binomial distribution to derive the probability of finding exactly r plants in a randomly chosen quadrat of area 1.0 m^2.
(ii) Calculate the probabilities, $p(r)$, for $r = 0, 1, 2, 3, 4, 5, 6$.
(iii) Plot the results as a histogram.

(i) In this question, each of ($n =$) 1500 plants is a statistical *trial*, and may be either inside (Y) or outside (N) the specific quadrat area of 1.0 m^2.

The value of the probability, p, is the probability that a *particular* plant will appear in the quadrat. This will be given by the area of the quadrat divided by the total area:

$$p = 1.0/1000 = 0.001$$

The values for $p(r)$ can then be calculated using:

$$p(r) = {}_{1500}C_r \times 0.001^r \times (1 - 0.001)^{(1500-r)}$$

(ii), (iii) The specific values for $p(r)$ are calculated in the following table and plotted in Figure 8.8.

Number, r	0	1	2	3	4	5	6
Probability (binomial), $p(r)$	0.223	0.335	0.251	0.126	0.047	0.014	0.004

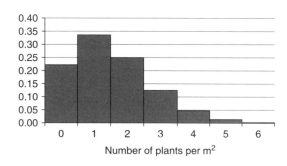

Figure 8.8   Probabilities of finding specific numbers of plants.

Note that we will see in Example 8.12 that this type of problem, with a low value for p, approximates to the Poisson distribution.

Q8.12

A certain drug treatment cures 70 % of people having a particular disease. If ten people suffering from the disease are treated, calculate the probability (to 3 significant figures) that:

  (i) no one will be cured;
  (ii) only one person will be cured;
 (iii) exactly eight people will be cured;
 (iv) exactly nine people will be cured;
  (v) exactly ten people will be cured;
 (vi) at least eight people will be cured;
(vii) no more than seven people will be cured.

8.4.3 Parameters of the binomial distribution

The **mean value**, μ, of the binomial distribution is given by:

$$\mu = n \times p \qquad\qquad [8.18]$$

When making experimental measurements of a binomial distribution, the *experimental mean value*, $\bar{r}$, would be a *best estimate* of the true mean, μ. The *best estimate* of the individual probability, p, would then be given by:

$$p \approx \frac{\bar{r}}{n} \qquad\qquad [8.19]$$

The **standard deviation**, σ, of the binomial distribution is given by:

$$\sigma = \sqrt{n \times p \times (1 - p)} \qquad\qquad [8.20]$$

Example 8.10

Calculate the probabilities, $p(r)$, of recording r heads when a balanced coin is tossed 100 times.

(i) Plot the results on a probability distribution as a function of r.
(ii) Calculate the mean and standard deviation of this distribution.

The worked answer is given in the following text.

In Example 8.10, the probability that a single toss of the coin will give a 'head' is $p = 0.5$. Using equation [8.17], the probability distribution will then be defined by:

$$p(r) = {}_{100}C_r \times 0.5^r \times 0.5^{(100-r)} \Rightarrow {}_{100}C_r \times 0.5^{100}$$

and is plotted in Figure 8.9.

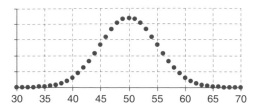

Figure 8.9   Relative probabilities of recording r heads from 100 throws.

Using equation [8.18], the mean value of the distribution in Figure 8.9 is:

$$\mu = n \times p \Rightarrow 100 \times 0.5 \Rightarrow 50$$

Using equation [8.20], the standard deviation of the distribution in Figure 8.9 is:

$$\sigma = \sqrt{n \times p \times (1 - p)} \Rightarrow \sqrt{100 \times 0.5 \times 0.5} \Rightarrow \sqrt{25} \Rightarrow 5$$

The distribution in Figure 8.9 is symmetrical, and we will see (8.4.7) that it can be approximated to a normal distribution. However, if $p \neq 0.5$, the distribution can be very non-symmetrical as in Figure 8.8.

Example 8.11

The aim of this exercise is to use the results for $p(r)$ from Example 8.9 and use the techniques from 7.3.6 to estimate the mean value, $\bar{r}$, of the number of plants found per square metre.

Using the probabilities, $p(r)$, for $r = 0, 1, 2, 3, 4, 5, 6$:

(i) Calculate the sum of the probabilities over this range, $\sum_r p(r)$.
(ii) Calculate the products, $r \times p(r)$, for each value of r, and the sum of these values, $\sum_r r \times p(r)$.
(iii) Calculate the mean value of $r, \bar{r} = [\sum_r r \times p(r)]/[\sum_r p(r)]$.

The calculations for (i) and (ii) are completed in the following table:

Number, r	0	1	2	3	4	5	6	Total
Probability (binomial), $p(r)$	0.223	0.335	0.251	0.126	0.047	0.014	0.004	0.999
$r \times p(r)$	0.000	0.335	0.502	0.377	0.188	0.070	0.021	1.495

Note that, in the table above, the sum of probabilities, $\sum_r p(r)$, is not exactly equal to 1.000. This is because there is still a very small probability that seven or more plants may be recorded, and we have not included this in our total.

The experimental mean value can be calculated:

$$\bar{r} = \left[\sum_r r \times p(r)\right] \Big/ \left[\sum_r p(r)\right] = 1.495/0.999 \approx 1.50$$

The true mean of the distribution is given by equation [8.18]:

$$\mu = n \times p \Rightarrow 1500 \times 0.001 \Rightarrow 1.5$$

which agrees with the result calculated from the values of $p(r)$.

8.4.4 Poisson distribution

As with the binomial distribution, the Poisson distribution calculates the probabilities of observing a certain number, r, of outcomes, each of which has a probability, p, of occurring. However, the Poisson distribution assumes that the probability that an individual trial will produce a specific outcome is very small, i.e. $p \ll 0.1$.

Typical examples of Poisson distributions include situations where relatively few events are being counted in a small section of a much larger environment, such as:

- numbers of plants in a small area of a large field;
- radioactive decay events occurring in a small time interval out of an overall decay period;
- occurrences of a rare illness in a town.

In each of the above examples, the probability, p, that a particular event (plant, decay event, ill person) will be found in the particular situation (area, time interval, town) will be small.

The Poisson probability, $p(r)$, of finding r occurrences of a particular outcome is defined by:

$$p(r) = \frac{e^{-\mu} \times \mu^r}{r!} \qquad [8.21]$$

where μ is the *mean number* of such outcomes that would normally be expected to occur.

Note that the Poisson probability does not depend on a value of n (total number of possible trials).

The Excel function POISSON(r, μ, False) can be used to calculate values for the probabilities of a binomial distribution. Entering 'True', instead of 'False', in the last argument value would give the *cumulative* probability of $p(r)$.

We now use Example 8.12, in comparison with Example 8.9, to investigate the use of the Poisson equation when p is very small.

Example 8.12

It is known that, on average, a random spread of particular rare plants in a field gives an average of 1.5 plants per square metre. Use the Poisson distribution to derive the probability of finding exactly r plants in a quadrat of area 1.0 m^2.

Calculate the probabilities, $p(r)$, for $r = 0, 1, 2, 3, 4, 5, 6$. (Note the similarity with the question in Example 8.9.)

The worked answer is given in the following text.

In Example 8.12, we are given the mean value, $\mu = 1.5$, of the number of plants that will fall within an area of 1.0 m^2.

As we are testing an area of 1.0 m^2 out of a whole field, then the probability, p, that any particular plant will randomly fall within the quadrat will be very small. For example, the value of p in Example 8.10 was calculated to be 0.001. Hence we can use the Poisson probability from equation [8.21]:

$$p(r) = \frac{e^{-1.5} \times 1.5^r}{r!}$$

The results of the calculation for $r = 0, 1, 2, 3, 4, 5, 6$ are given in Table 8.3, together with the results from the binomial calculation from Example 8.9.

Table 8.3.

Number, r	0	1	2	3	4	5	6
Probability (Poisson), $p(r)$	0.223	0.335	0.251	0.126	0.047	0.014	0.004
Probability (binomial), $p(r)$	0.223	0.335	0.251	0.126	0.047	0.014	0.004

It can be seen from Table 8.3 that, to 3 decimal places, the results from the Poisson and binomial distributions are the same for this problem with $p = 0.001$.

Q8.13

The number of deaths due to a specific disease per year per 100 000 people is on average 50. Assuming that the occurrence of each incidence of the disease is random, calculate the probability that in a particular group of 1000 people there may be:

(i) no deaths   (iii) two deaths
(ii) one death   (iv) more than two deaths

8.4.5 Using binomial and/or Poisson distributions

When the individual probability, p, is very small, then it is possible to use either the binomial or the Poisson distribution. However:

- for the binomial distribution it is necessary to know values for n and p; but
- for the Poisson distribution it is only necessary to know the value of μ.

It is useful to note that, for the binomial distribution, the maximum value for r would be equal to n, but that for the Poisson distribution there is no theoretical maximum value for r.

The use of the above ideas is illustrated in Example 8.13.

Example 8.13

A service engineer is responsible for responding to emergency breakdowns from 2500 customers. He normally receives four calls per day.

Calculate the probability that he would receive eight calls in a particular day.

(assume no customer has more than one breakdown per day)

Use *both* the binomial and Poisson distributions to answer the question.

This is a binomial-type problem because any one customer will only have a breakdown (Y) or no breakdown (N).

Calculation assuming a binomial distribution

We need to calculate values for n and p to use in [8.17].

The value of n equals the maximum possible value for r, which is the maximum number of calls per day, $n = 2500$.

The mean number of calls is known to be $\mu = 4$.

From [8.18], the probability, p, that any one customer will have a breakdown, is:

$$p - \mu/n = 4/2500 = 0.0016$$

We need to calculate $p(r)$, where $r = 8$:

$$p(8) = {}_{2500}C_8 \times 0.0016^8 \times (1 - 0.0016)^{2492} = 0.02972$$

Calculation assuming a Poisson distribution

For the Poisson distribution we already have the mean value, $\mu = 4$. Hence:

$$p(8) = \frac{e^{-4} \times 4^8}{8!} = 0.02977$$

The two results are virtually identical.

8.4.6 Cumulative distributions

The expression for $p(r)$ calculates the probability that exactly r occurrences of a particular condition (Y) will occur. In practice, many problems are concerned with the *combined* probability of several possible values of r occurring.

The cumulative probability (see 7.3.9) for all values of r from 0 to r' is given by:

$$cp(r \leqslant r') = p(0) + p(1) + \cdots + p(r' - 1) + p(r')$$

Example 8.14

Using the data from Table 8.3, we can calculate the probabilities of finding:

(i) two or less plants per quadrat
(ii) three or more plants per quadrat.

Here:

(i) $cp(r \leqslant 2) = p(0) + p(1) + p(2) = 0.223 + 0.335 + 0.251 = 0.809$
(ii) $cp(r \geqslant 3) = 1 - cp(r \leqslant 2) = 1 - 0.809 = 0.191$

8.4.7 Approximations for the binomial distribution

There are two important approximations for the binomial distribution, which occur when:

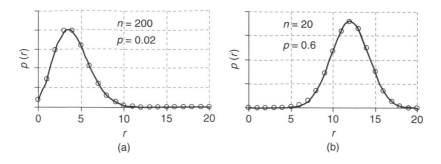

Figure 8.10 Comparisons of binomial distribution (circle points).

- $p \ll 1$ (i.e. p is very small)

  the probabilities of the binomial distribution become approximately the same as the *Poisson distribution*.

  Figure 8.10(a) compares the binomial distribution ($n = 200$, $p = 0.02$) with the Poisson distribution ($\mu = n \times p = 4$) shown as a solid line.

- $np(1 - p) \gg 1$ (typically $np(1 - p) \geqslant 5$)

  the probabilities of the binomial distribution become approximately the same as the *normal distribution*. Figure 8.10(b) compares the binomial distribution ($n = 20$, $p = 0.6$) with the normal distribution ($\mu = n \times p = 12$, $\sigma = \sqrt{np(1 - p)} = 2.19$) shown as a solid line.

The approximation, $p \ll 1$, to the Poisson distribution is considered in 8.4.4. The approximation, $np(1 - p) \gg 1$, to the normal distribution is useful because we can then use the equations developed for the normal distribution. For example, it is possible to develop the confidence interval, *CI* (at *X*%), for the true mean of the binomial distribution (see also 8.2.4):

$$\text{Confidence interval, } CI_{X\%} \approx \bar{r} \pm \left[t_{2,\alpha,\infty} \times \sqrt{np(1 - p)} \right] \approx \bar{r} \pm \left[t_{2,\alpha,\infty} \times \sqrt{\bar{r} \times \left(1 - \frac{\bar{r}}{n} \right)} \right]$$

$$[8.22]$$

where $\bar{r}$ is the experimentally measured value of r, and $t_{2,\alpha,\infty}$ is the 2-tailed t-value with $df = \infty$ (e.g. $t_{2,0.05,\infty} = 1.96$).

The one proportion test, given in 14.3.2, is the hypothesis test equivalent to the above confidence interval.

Example 8.15

A country is due to have a referendum with only two choices – Yes or No. In anticipation of the referendum, 1000 people are selected at random in an 'opinion poll' and it is found that 520 will vote Yes and 480 will vote No.

Calculate the 95 % confidence interval for the true proportion of people who will vote Yes.

From the sample, $n = 1000$ and $\bar{r} = 520$.

The two-tailed confidence interval (see [8.22]) for the true number of Yes votes out of each group of 1000 people will be:

$$CI_{95\%} \approx 520 \pm \left[1.96 \times \sqrt{520 \times \left(1 - \frac{520}{1000} \right)} \right] \Rightarrow 520 \pm 31$$

Thus, for each group of 1000 voters, the number predicted (at 95 % confidence) to vote Yes lies between 489 and 551.

This converts to a proportion of between 0.489 and 0.551.

Q8.14

A random sample of 50 frogs is taken from a lake, and it is found that 37 are female and 13 are male. Estimate, with a 95 % confidence interval, the true proportion of female frogs in the lake's frog population.

(See also Example 14.6)

8.4.8 Shape of a distribution

The mean and standard deviation of a distribution are, respectively, measures of location (or position) and spread (or width).

As we have seen (compare Figures 8.8 and 8.9) distributions can also have different *shapes*. Other statistical parameters have been introduced to describe different shape characteristics:

Skewness. is a measure of the extent to which the distribution is skewed to one side or the other.

Kurtosis. is a measure of the extent to which the top of the distribution may be either flatter or more peaked than the normal distribution.

A more in-depth discussion of these is given on the Website.

One further measure of interest is the ratio of the variance of the distribution to the mean value. The **coefficient of dispersion**, **CD**, is given by:

$$CD = \frac{\text{Variance}}{\text{Mean}} = \frac{\sigma^2}{\mu}$$

which, using equations [8.18] and [8.20], for the binomial distribution becomes:

$$CD = \frac{n \times p \times q}{n \times p} = q \qquad [8.23]$$

A good example in the use of the coefficient of dispersion, CD, is in the distribution of random 'events', e.g. plants in a field:

- If the events (plants) are individually distributed at random, then the number of plants in a small area will be given by the Poisson distribution (see Example 8.12). For the Poisson distribution, $CD \approx 1$.
- If there is a supportive 'interaction' between the events (plants), then the events may group together (plant clumping). Another example would be the development of 'clusters' of an otherwise random disease, e.g. possible increase in the incidence of disease cases due to local pollution. In this case, the increased probability of 'high-count' groups would give $CD > 1$.
- If the events compete for resources, then they will tend to spread out and give a more even distribution. This will give a smaller dispersion of values around the mean value, resulting in $CD < 1$.

Q8.15

In assessing the distribution of a species of plant, the two data sets, A and B, in the table below record the frequencies with which specific numbers were recorded in 100 quadrats. The means and standard deviations of the two distributions are given.

Number of plants	0	1	2	3	4	5	6	7	8	9	10	Mean	Standard deviation
Frequency, A	5	13	21	22	18	11	6	3	1	0	0	3.13	1.76
Frequency, B	10	14	17	18	16	12	7	4	1	1	0	3.13	2.04

The frequency distributions are also recorded in the line graph shown in Figure 8.11.

(i) Calculate the coefficients of dispersion for the two data sets.

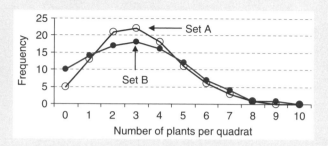

Figure 8.11 Plant distributions.

(ii) On the basis of (i), decide which data set shows a Poisson distribution.
(iii) For the data set that is not Poisson, decide whether the distribution of plants shows possible clumping of plants or a more even spread.

9

Scientific Investigation

Overview

 • Excel files appropriate to selected Examples.
• Example hypothesis tests in Excel and Minitab.

Science is concerned with understanding the way in which the natural world works. This includes the biology of the living world, the chemistry of atomic and molecular structures, the physics of forces and systems, but it also includes the more applied sciences, e.g. sports science, forensic science, the environmental sciences, etc.

In this chapter we see that science is used to investigate a wide range of different systems, and that there are many different types of 'scientific investigation'.

We also see that the power and effectiveness of science, as a way of accurately describing nature, grew out of the Renaissance in the sixteenth century with the development of the 'scientific method'.

As an essential feature of the 'scientific method', we concentrate on the issues involved with the setting and testing of hypotheses. The statistical implementation of hypothesis tests for a range of different system parameters is developed using both parametric and non-parametric tests.

The final chapter in the book takes hypothesis testing into more complex scientific investigations, and highlights the interdependence between the experiment design and the statistics involved in assessing the validity of the proposed hypotheses.

9.1 Scientific Systems

Any system that is being investigated can normally be characterized by:

- **Outcomes** – observable changes in the system.
- **Factors** – 'inputs' to the system that may affect the outcomes.
- **Mechanisms** – the actual processes (often hidden) within the system that link the 'input' factors to the 'observed' outcomes.
- **Subjects** – representative examples of the 'scientific system' that are used for the particular experiment.

Essential Mathematics and Statistics for Science 2nd Edition   Graham Currell and Antony Dowman
Copyright © 2009 John Wiley & Sons, Ltd

Each 'factor' will have two, or more, 'levels', and a change of 'level' may cause a particular 'outcome' from the system. 'Outcomes' and 'levels' will be measured by suitable experimental variables.

Example 9.1

It is claimed that omega 3 oils, found in many fish, reduce 'clogging' of the arteries. In this case the 'system' is the human body, the 'outcome' is reduced deposits in the arteries, the 'factor' relates to different 'levels' of intake of omega 3 oils, and the 'mechanism' is the way in which the body metabolizes these oils. In an experiment, suitable 'variables' must be used to measure the amount of arterial deposit and the intake ('levels') of omega 3 oils.

There are three main levels in which science can normally understand such a system, such as that described in Example 9.1. It may be possible to:

1. Confirm a **correlation (or association)** between the input factor(s) and the system outcome(s), e.g. an outcome may change following a change in a factor. However, a correlation between factor and outcome does *not necessarily* mean that the change in the factor is the *reason* for the change in the outcome.
2. Confirm a **causal relationship** between the input factor(s) and the system outcome(s). In this case, a change in a factor (cause) can be demonstrated to be the *reason* for the change in outcome (effect).
3. Explain the **mechanism** by which the factor causes the outcome to change. The understanding of this mechanism must give more than a description of *past* observations; it must be sufficiently good to be able to *predict* how the system will work under modified conditions.

In the case of the omega 3 example (Example 9.1):

- At level 1, the variable that records arterial deposits and the variable that measures oil intake may show close correlation, but the oil intake might not be the 'cause' of the reduction in deposits. For example, the reduction in arterial deposits may be due to the fish diet for reasons not related to the omega 3 oils. The fish diet may be the 'cause' that both reduces arterial deposits (an effect) and, coincidentally, increases the intake of omega 3 oils (now an effect and not a cause).
- Alternatively at level 2, a 'cause and effect' relationship could be shown to exist between 'factor' and 'outcome' (i.e. the omega 3 oils actually reduce arterial deposits), but the mechanism by which this works might not be fully understood.
- Ideally, at level 3, science should be able to demonstrate, within the constraints of the 'scientific method' (9.2), a robust understanding of the mechanism involved in the process. This understanding must continue to be essentially correct as further discoveries are made about the system.

Scientific 'systems' can be recognized in all branches of science – from an understanding of the properties of subatomic particles to the behaviour of the Universe itself. In all cases, the development of understanding identifies the same three levels: *correlation, causation, mechanism*.

9.2   The 'Scientific Method'

At a minimum level, any work in science must be characterized by an accuracy and objectivity in the collection of data, and a commitment to base conclusions solely on experimental results.

An important distinction should now be made between the processes used to reach conclusions:

Induction: *comparative* conclusions about one system are made on the basis of what is known to happen in a *different* system.
Deduction: *new* conclusions are drawn as direct *logical* consequences of known facts.

Inductive reasoning may be considered to be equivalent to an *intelligent guess* that might prove to be wrong. Inductive reasoning is often a good starting point to consider how a new system might work and suggest possible hypotheses, but we will see that deduction, and not induction, is the fabric of true science.

Example 9.2

In trying to understand the mechanism of elasticity in the common elastic band, we may compare the effect with elasticity in other materials. For example, we know that when we pull on a metal wire it will stretch and when we pull on an elastic band it will also stretch. We may make an *induction* that the mechanisms are basically similar in the two cases, although different in magnitude.

If our *induction* is correct, we might expect that, when we heat up an elastic band, it will stretch in the same way that a metal wire will stretch on heating. However, when we perform the experiment, we find that a stretched elastic band will tend to *contract* on heating. Our *induction* about similar mechanisms must be wrong in this case!

Deductive reasoning starts from known facts, and then proceeds to work out the inevitable consequences of those facts. The process of deduction is faultless, but the accuracy of the conclusions still depends on the accuracy of the original facts on which the deduction was based.

The 'scientific method' itself is based on a cycle of hypothesis, experiment, deduction and prediction:

- *Observations* or *inductive reasoning* are often the starting point where observations begin to suggest a possible **mechanism** that might explain the behaviour of a system. (In Example 9.2 it is thought that the mechanism for elasticity in the elastic band could be similar to that in the metal wire.)
- A **hypothesis** is developed which describes an aspect of the proposed mechanism in the scientific system. The choice between the truth or otherwise of this hypothesis is often written as two alternatives:
  H_1: Proposed hypothesis – the proposed mechanism *is* correct.
  H_0: Null hypothesis – the proposed mechanism is *not* correct.
  (In Example 9.2 the proposed hypothesis is that the mechanism in the elastic band is similar to that in the metal wire.)

- An **experiment** is devised, whose outcome should give evidence to support, or not support, the proposed hypothesis, H_1. (In Example 9.2 the proposed hypothesis predicts that an elastic band should expand on heating.)
- The experiment is performed and a decision is made as to whether the new understanding of the system is likely to be true. The conclusion is a **deduction** based on the observed facts. (In Example 9.2, the observed facts do *not* support the proposed hypothesis. A new mechanism must be found to explain the behaviour of the elastic band.)
- If the proposed hypothesis is supported, then the proposed mechanism should be used to *predict* other observable behaviour patterns for the system. If these predictions are tested using the 'hypothesis-testing' approach, and continue to confirm the new explanations, then this understanding begins to be accepted as a good description of the system.

The process of the 'scientific method' is also referred to as the 'hypothetico-deductive' method, emphasizing the two key elements of the concept.

When a hypothesis continues to satisfy more and more experimental tests, then it will gradually be accepted as a scientific fact. However, it must be remembered that the process of the 'scientific method' does not end with the discovery of 'fact', just because it satisfies all current experimental data. Any scientific 'fact' may be modified in the future if further experimental results fail to conform to the accepted theory.

Example 9.3

An excellent example of the 'scientific method' occurred in the development of the theory of relativity by Albert Einstein. The new theory, for which there were many doubters, predicted that light from a star would be *deflected* as it passed close to our Sun.

This prediction would test the theory of relativity, and in 1919 during an eclipse of the Sun the apparent positions of stars close to the Sun were found to shift according to Einstein's prediction. Since that time, the theory of relativity has continued to predict events accurately, and is now an accepted 'fact' of science.

However, as part of the 'cycle' of the scientific method, it is always probable that the current theory of relativity will be superceded by new 'facts' as more precise measurements demand a revised or alternative explanation.

9.3   Decision Making with Statistics

9.3.1   Introduction

Many investigations in science seek to answer a simple 'Yes or No' question. For example:

- Is a new vaccine effective in preventing an infection?
- Do glass fragments at the scene of a crime come from different sources?
- Does a particular training regime improve athletic performance?

The statement of the initial 'Yes or No' question is a hypothesis statement, which we can present in two parts:

- 'Yes' is the Proposed hypothesis; and
- 'No' is the Null hypothesis.

However, any scientific investigation is subject to errors and uncertainties, and the results do not give an *absolute* answer to the 'Yes or No' question – this is where the statistics come in! The statistical analysis in a hypothesis test is based on the experimental data recorded, and normally calculates the probability (p-value) that it would be wrong to choose the 'Yes' (proposed hypothesis) answer to the question. There is never an absolute certainty in the final answer, but at least it is possible to assess the probability of making at least one type of error.

We will introduce hypothesis testing by working through an extended example in 9.3.2. The following sections then highlight the separate elements of the hypothesis-testing procedure.

9.3.2 Calculation of p-values

We will introduce the concept of the p-value in hypothesis testing by using the investigation outlined in Example 9.4.

Example 9.4

A sports club follows a long-standing tradition in which it always tosses the same special coin at the start of a game. However, some club members believe that the coin might be biased so that it produces more heads than tails. They ask you to toss the coin 100 times and then, using your knowledge of statistics, decide whether the coin is indeed biased in favour of heads.

The analysis is performed in the following text.

You accept the task outlined in Example 9.4. You realize that the true situation must be one of two alternatives, which you express in terms of a proposed hypothesis (labelled as H_1) and its 'null' option (labelled as H_0):

Proposed hypothesis, H_1:	The coin gives more heads than tails.
Null hypothesis, H_0:	The coin does not give more heads than tails.

You will toss the coin $n = 100$ times, and on the basis of the number of heads, r, recorded, you will decide whether you will 'accept' or 'not accept' that the proposed hypothesis, H_1, is correct.

Clearly there could be two possible **correct** outcomes. You might be correct by 'Accepting H_1' for a biased coin, or you might be correct by 'Not accepting H_1' where there is no bias.

However, there could be two different types of **error** that you could make, each with different consequences:

- **Type I error.** You accept H_1 believing that the coin is biased, when in fact it is perfectly fair. The consequence is that you have unnecessarily ended the long-standing tradition.
- **Type II error.** With limited experimental data, you choose *not* to accept H_1, but the coin is actually biased. Your poor experimental skills have failed to identify a real problem.

The four possibilities are illustrated in Table 9.1. The *columns* give the 'true' condition of the coin, and the *rows* give your decision.

You realize that, due to the random nature of the process, there will be an inherent uncertainty when you make your decision. You decide that you will only claim that the proposed hypothesis, H_1, is true if the probability of error is less than (or equal to) 5%.

Table 9.1.

Your decision:	True condition of the coin	
	H_1 is true	H_1 is not true
Accept H_1	**Correct** The coin is replaced Well done!	**Type I error** You unnecessarily end an important tradition.
Do not accept H_1	**Type II error** Your experimental skills are inadequate	**Correct** No action is necessary Well done!

In general, we define the **significance level,** α (alpha), as the *largest probability of error* that is acceptable when choosing the proposed hypothesis, H_1. For this experiment, you have chosen a decision criterion of $\alpha = 0.05$. (Note that a 5% probability is equal to a probability of 0.05 or '1 in 20'.)

You toss the coin **100 times**, expecting (for an unbiased coin) to record about 50 heads, but you actually record **60 heads**.

At first sight, it might appear that the coin *is* giving more heads than tails, but you decide to work out the probability, p, that an *unbiased* coin could, by random chance, record 60 or more heads.

The probabilities with which a *balanced* coin will produce different numbers of heads can be calculated using the binomial distribution (see Example 8.10), giving the probability distribution in the graph shown in Figure 9.1.

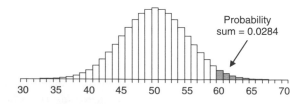

Figure 9.1   Probabilities that a *balanced* coin will record r heads in 100 throws.

From this distribution, it is possible to identify the key probability area:

Probability of recording 60 heads or more, $p(r \geqslant 60) = 0.0284$

You now realize that the p-value, $p = 0.0284$, is:

- the probability that a *perfectly balanced* coin might, by random chance, record 60 or more heads;
- the probability that the null hypothesis, H_0, might be true;
- the probability of a Type I error if you accept H_1 (reject H_0).

On the basis of the information in the text, you must now make the critical decision, comparing the observed probability p-value, $p = \mathbf{0.0284}$, with your initial significance level criterion, $\alpha = \mathbf{0.05}$:

Do you '**Accept H_1**' and declare the coin to be biased, or do you '**Not accept H_1**' and say that you cannot be sure that it is biased?

As the probability that you will be wrong $(p = 0.0284)$ is *less than* your upper limit of $(\alpha =) 0.05$, you actually choose to accept H_1 and claim that coin is biased.

The old coin is discarded and the club committee choose a new coin.

Example 9.5

Having discarded the original coin (Example 9.4), the committee now ask you to check whether the *new coin* might be biased, giving *either* more heads *or* more tails.

The worked answer is given in the following text.

The question posed in Example 9.5 requires a new experiment, so you establish the new hypotheses which take into account that the coin might be biased in *either* direction:

Proposed hypothesis, H_1:	The number of heads is different from the number of tails.
Null hypothesis, H_0:	The number of heads is the same as the number of tails.

In this case you would accept that the coin is biased (H_1) if it gives *either* a significantly high *or* a low number of heads.

You toss the new coin 100 times and again you record 60 heads. However, you now express this result as:

Difference between the recorded result of 60 and the 'expected' result of $50 = \mathbf{10}$.

You now need to work out the probability, p, that an unbiased coin could, by random chance, record a *difference* of 10 or more heads, i.e. either 40 or less, or 60 or more.

The probabilities with which a *balanced* coin will produce different numbers of heads can be calculated again using the binomial distribution (Figure 9.2).

From this distribution, it is possible to identify the total probability area of recording a *difference* or 10 or more heads:

$$p(\text{difference} \geqslant 10) = p(r \leqslant 40) + p(r \geqslant 60) = 0.0284 + 0.0284 = 0.0568$$

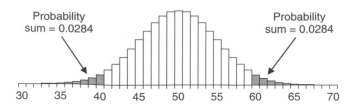

Figure 9.2   Probabilities that a *balanced* coin will record *r* heads in 100 throws.

You now realize that the *p*-value, $p = 0.0568$, is:

- the probability that a perfectly balanced coin might, by random chance, record either 40 or less, or 60 or more heads;
- the probability that the null hypothesis, H_0, might be true;
- the probability of a Type I error if you accept H_1 (reject H_0).

On the basis of the information in the text, you must now make the critical decision, comparing the observed probability *p*-value, **$p = 0.0568$**, with your initial significance level criterion, $\alpha = \mathbf{0.05}$:

> Do you '**Accept H_1**' and declare the coin to be biased, or do you '**Not accept H_1**' and say that you cannot be sure that it is biased?

As the probability that you will be wrong ($p = 0.0568$) is *greater than* your upper limit of ($\alpha =$) 0.05, you do not accept H_1 because there is not enough evidence to say that this coin is biased.

There might appear to be an apparent contradiction, because you accepted H_1 for the first test but did not accept H_1 for the second test even though you had obtained the *same* experimental results (60 heads) for both tests. However, there is no contradiction, because the different conclusions relate to different *questions*.

You *must decide, before you perform the experiment*, whether you are just testing for a *difference* between an observed and expected value, or whether you are testing that one value is *greater* (or *less*) than the other. If you change your mind after performing the experiment, then you must conduct the experiment again to collect *new* data. You must not look at the data *before* deciding which type of test to apply on that *same* data – this may lead to an invalid conclusion.

9.4   Hypothesis Testing

9.4.1   Introduction

This unit provides a concise review of the main elements involved in the hypothesis-testing process. However, to gain a good understanding of the underlying processes involved in hypothesis testing, it is recommended that readers work through Examples 9.4 and 9.5 in the previous unit entitled 'Decision Making with Statistics'.

9.4.2 Hypothesis statement

It is essential, *before* performing any experimental measurements, to *state the proposed hypothesis* being tested, usually written as H_1, and the *null hypothesis*, usually written as H_0:

Proposed hypothesis, H_1:   Smoking increases the probability of lung cancer.
Null hypothesis, H_0:   Smoking does not affect the probability of lung cancer.

The proposed hypothesis usually demonstrates that a factor (9.1) does indeed cause some 'effect' or 'change of value'. The null hypothesis usually demonstrates 'no factor effect' or the 'status quo'.

A scientific experiment must be precise about what is actually being measured, and will normally measure some physical, chemical or biological variable, which will then be used as an *indicator* in choosing which hypothesis is believed to be correct.

It is also important that any 'vague' terminology should be accurately defined – for example, 'smoking' in the above example may be defined as five cigarettes per day.

Example 9.6

As part of a monitoring procedure, an analyst wishes to test whether the lead content of a water supply *exceeds* 50 ppb. State the relevant hypotheses.

If the true lead content is μ, the appropriate hypotheses for this test would be:

Proposed hypothesis, H_1:   Lead content is greater than 50 ppb, $\mu > 50$ ppb.
Null hypothesis, H_0:   Lead content equals 50 ppb, $\mu = 50$ ppb.

9.4.3 Tails

In defining a hypothesis statement in respect of a possible factor 'effect', there are two possibilities:

- **2-tailed:** the factor effect may operate in *either direction*; or
- **1-tailed:** the factor is only being tested for an effect in *one particular direction*.

The hypothesis statement in Example 9.6 is '1-tailed', because the proposed hypothesis only proposes an *increase* in lead concentration. Example 9.7 gives an example of a '2-tailed' hypothesis in a similar context. In this case the proposed hypothesis proposes *either* an increase *or* a decrease.

Example 9.7

As part of a monitoring procedure, an analyst wishes to test whether the lead content of a water supply is *not equal* to 50 ppb. State the relevant hypotheses.

If the true lead content is μ, the appropriate hypotheses for this test would be:

Proposed hypothesis, H_1:   Lead content does not equal 50 ppb, $\mu \neq 50$ ppb.
Null hypothesis, H_0:   Lead content equals 50 ppb, $\mu = 50$ ppb.

It is useful to remember that, for the same data set (for a symmetrical distribution) :

$$p\text{-value for a } \textit{2-tailed} \text{ test} = 2 \times p\text{-value for a } \textit{1-tailed} \text{ test} \qquad [9.1]$$

Q9.1

Decide, for each of the following statements for proposed hypotheses, which implies a 1-tailed test and which implies a 2-tailed test.

	Number of tails
(i) An athlete's performance is enhanced when using a particular dietary supplement.	1/2
(ii) Low-level background music affects people's capacity to memorize a list of names.	1/2
(iii) The mean content of breakfast cereal packets filled by a certain filling machine is less than 580 g.	1/2
(iv) An increase in a car driver's blood–alcohol level to 40 mg per 100 mL will change the probability of having an accident.	1/2

9.4.4 Decisions based on the *p*-value

The decision process starts with the assumption that the null hypothesis might be true.

Statistics software is used to calculate the probability (**p-value**) that the null hypothesis could produce the observed experimental results. If this probability is sufficiently *small* (i.e. it is unlikely that the null hypothesis is true), then the proposed hypothesis would be accepted as being true instead.

With this type of question, there are two ways of being *wrong*:

- deciding 'proposed' when 'null' is correct (Type I error); and
- deciding 'null' when 'proposed' is correct (Type II error).

In terms of the hypotheses, the Type I and Type II errors could occur as in Table 9.2.

Table 9.2.   Errors in a hypothesis test.

Conclusion:	True situation	
	H$_1$ true	H$_1$ not true
Accept H$_1$	Correct result	Type I error
Do not accept H$_1$	Type II error	Correct result

In making the decision that the proposed hypothesis is true, it is necessary to decide *how small* the *p*-value must be so that we can confidently reject the null hypothesis and avoid a Type I Error.

Before starting the experiment, we must choose the value of the significance level, α (see 9.3.2). The **significance level,** α (alpha), is the *largest probability of error* that is acceptable when choosing the proposed hypothesis, H$_1$.

If a conclusion is claimed to be *significant*, then the results normally indicate that H$_1$ is correct with less than a 0.05 (5 %) chance of being wrong. If a conclusion is claimed to be *highly significant*, then the probability of H$_1$ being wrong is normally less than 0.01 (or 1 %).

Note that the term **confidence level** is often used to express the percentage probability of being *right*, e.g. confidence levels of 95 % and 99 % for being right are equivalent to significance levels of 0.05 and 0.01, respectively, for being wrong!

The choice of a value for the significance level, α, depends on the context of the problem, and the consequences that would arise if H$_1$ were chosen in error. The default level for most scientific research is taken as $\alpha = 0.05$.

The *probability* of a Type I error for a *given experiment* is given by the *p*-value, which can be calculated from the experimental results – see 9.3.2. The *probability* of a Type II error is denoted by β (beta), which is usually much more difficult to calculate.

The significance level, α, is the maximum probability of a Type I error that would be acceptable. Its value must be decided as part of the *initial* experiment design process.

On the basis of the relative values of α and p, it is then possible to make the decision:

If $p \leqslant \alpha$ Accept proposed hypothesis, H$_1$, at a significance level of α

[9.2]

If $p > \alpha$ Insufficient evidence to accept proposed hypothesis

---

Most software programs (e.g. Excel, Minitab) will calculate a *p*-value for the given statistical test. This makes performing a statistical test very easy, as the decision is based only on the relative sizes of p and α.

9.4.5   Power of the experiment

The **power** of an experiment is a measure of how successful the experiment will be in confirming the existence of a 'factor' effect, if such an effect actually exists. It is the 'opposite' of the probability of a Type II error:

$$\text{Power} = 1 - \beta \qquad\qquad [9.3]$$

Ideally the power of the experiment should be as close to 1.0 as possible. In practice, the power of an experiment is often limited by many experimental uncertainties that may make the existence of the 'factor' effect difficult to detect.

It is only possible to calculate the value of *power* (and β) for a particular experiment if all the effects of any random uncertainties and external bias are known and understood. This is not normally the case.

9.4.6   Stating results

Before choosing the proposed hypothesis, H_1, it is necessary to define the significance level for making the decision. Hence, an *appropriate* statement of the 'Yes' conclusion would then be:

Accept the proposed hypothesis, H_1, at a significance level of (for example) 0.05.

However, it is *not* usually possible to calculate the probability of the Type II error for accepting the null hypothesis, H_0. In other words, it is often difficult to 'prove' a negative. Do *not* therefore use the statement 'Accept the null hypothesis'.

An *appropriate* statement for the 'No' result would therefore be either:

Insufficient evidence to accept the proposed hypothesis, H_1, at a significance level of (for example) 0.05

or:

Insufficient evidence to reject the null hypothesis, H_0, at a significance level of (for example) 0.05.

Q9.2

The results of a statistical test give a *p*-value, $p = 0.03$. Which of the following statements could be correct?

(i)   Accept the proposed hypothesis, H_1, at a significance   True/False
       level of 0.01.

(ii) Accept the proposed hypothesis, H_1, at a significance   True/False
       level of 0.05.

(iii) The null hypothesis, H_0, should be accepted at a   True/False
       significance level of 0.01.

(iv) Insufficient evidence to reject the null hypothesis, H_0,   True/False
       at a significance level of 0.05.

9.4.7 Statistic and critical value method

We have already demonstrated how a computer calculation of p-values provides the calculated probability value on which the outcome of the test is decided. However, an alternative method of calculation exists, which predates the use of computers. It is important to be aware of the general procedure involved when using **test statistics**:

1. For each test it is possible to calculate the value of a **test statistic** for the given data set. Each test will have its own statistic, e.g. t_{STAT} when testing mean values, χ^2_{STAT} when testing frequencies.
2. The value of the statistic depends on the strength of the experimental evidence that the hypothesis, H_1, might be true.
3. To make the decision on whether to accept the proposed hypothesis or not, it is necessary to compare the relative magnitudes of the calculated test statistic with a critical value (e.g. t_{CRIT}, χ^2_{CRIT}) that can be obtained from sets of tables.
4. The choice of the appropriate critical value for a given experiment will depend on the significance level, α, required and the degrees of freedom available in the experimental data – see the *individual tests* in Chapters 10 to 14.

Example 9.8

Two of the following sets of results from a t-test are valid. Two of the sets have inconsistent data values. In this particular t-test, the proposed hypothesis is accepted if $t_{STAT} \geqslant t_{CRIT}$.

Identify the **two** result sets that have inconsistencies:

A $t_{STAT} = 2.24$, $t_{CRIT} = 1.86$, $\alpha = 0.05$, $p = 0.028$
B $t_{STAT} = 2.46$, $t_{CRIT} = 2.68$, $\alpha = 0.05$, $p = 0.015$
C $t_{STAT} = 2.45$, $t_{CRIT} = 2.45$, $\alpha = 0.05$, $p = 0.02$
D $t_{STAT} = 2.61$, $t_{CRIT} = 3.00$, $\alpha = 0.01$, $p = 0.017$

Set A is consistent – both comparisons give the same conclusion:
$t_{STAT} > t_{CRIT}$ leads to 'Accept the proposed hypothesis'.
$p < \alpha$ leads to 'Accept the proposed hypothesis'.

Set B is *not* consistent – data gives conflicting conclusions:
$t_{STAT} < t_{CRIT}$ leads to 'Insufficient evidence to accept the proposed hypothesis'.
$p < \alpha$ leads to 'Accept the proposed hypothesis'.

Set C is *not* consistent:
$t_{STAT} = t_{CRIT}$ exactly on the choice boundary, and we would then expect that $p = \alpha$, but in the data $p < \alpha$.

Set D is consistent – both comparisons give the same conclusion:
$t_{STAT} < t_{CRIT}$ leads to 'Insufficient evidence to accept the proposed hypothesis'.
$p > \alpha$ leads to 'Insufficient evidence to accept the proposed hypothesis'.

9.5   Selecting Analyses and Tests

9.5.1   Introduction

This unit provides an introduction to some of the most commonly used statistical tests. For the detailed workings and examples of each test, readers are referred elsewhere in the book or on the Website.

An investigation in science often involves making one or more sets of replicate experimental measurements (7.1.2) which are statistical *samples* (7.2.2) of the *population* of all the measurements (possibly infinite) that could be made. A typical statistical test then calculates the probability (p-value) that the null hypothesis might explain the observed experimental data.

It is important to remember that the statistical test calculates the probability that the proposed hypothesis is *not* true. It does not directly calculate the probability that the proposed hypothesis *is* true.

The choice of an appropriate test is an integral part of the overall process of *designing the experimental procedure* – see Chapter 15. The detailed choice depends on a range of factors, but, in particular, it is essential to be clear about:

- what statistic is being tested, e.g. mean values, distribution of frequencies, proportions, etc.;
- whether the distribution of values being tested follows a specific distribution, e.g. a normal distribution;
- whether the test is 1-tailed or 2-tailed; and
- what is the significance level being used for the test?

9.5.2   Assumptions in using statistical tests

Statistical tests make certain assumptions about how the test was performed and possibly the underlying variation of the data being measured.

Each measurement must be *randomly selected* and *independent* of any other measurement.

For example, 'random selection' would require that seedlings being selected for different growth conditions *must* be chosen by an *objectively* random procedure (15.1.3).

The requirement for 'independence' between measurements aims to counteract any 'hidden' association between replicate measurements. This is particularly important in relation to possible errors in the measurement process itself. For example, if one instrument is used to measure one sample set of replicates and another instrument is used to measure a second sample set, then any bias between the instruments will be translated into an apparent difference between the experimental results.

There is a wide range of tests, called *parametric* tests, which make the assumption that the data values are being sampled from a known *distribution*, and for these tests the exact data values are used directly in the calculations.

An alternative range of tests, called *non-parametric* tests, do not make any assumption about an underlying distribution, and use data values only to *rank* the data in order.

The parametric t-tests, F-tests, ANOVAs, Pearson's correlation and regression all assume that the data is drawn from a *normal* distribution. This is often a valid assumption for most experimental data, particularly where mean values are calculated from large sample sizes. It has also been shown that these most commonly used tests are quite robust when dealing with data that is not normally distributed in that they still provide correct conclusions.

For some types of non-normal data, it is possible to apply a mathematical transformation to produce a distribution that is sufficiently close to normal (see the Website).

9.5.3 Common statistical tests

The first step is to be clear about what is being measured, e.g. mean values of two samples, proportions, frequencies, etc. This identifies the 'statistic' relevant to the particular test, leading to the following groups of tests:

Statistic:	Sample mean values (one or two samples)
Test:	Student's *t*-test
	Compares the *mean(s)* of one or two sets of replicate data Assumes that data is taken from a *normal* distribution
	One-sample *t*-test: Compares the mean of the sample data with a specific value. See Example 10.1
	Two-sample *t*-test: Compares the means of two data sample sets. See Example 10.5
	Paired *t*-test: Compares two sets of replicate data in which pairs of data items from each set have unique relationships. See Example 10.7

Statistic:	Sample mean values (multiple samples)
Test:	ANOVA – Analysis of Variance
	Compares the *mean(s)* of (usually) more than two sets of replicate data Assumes that data is taken from a *normal* distribution
	One-way ANOVA: Tests for the effect of *one factor* on the mean values of the samples. See Example 11.1
	Two-way ANOVA: Tests for the effects of *two factors* on the mean values of the samples. See Example 11.3 A two-way ANOVA *with replication* will also test for an *interaction* between the factors. See Example 11.6
	GLM – General Linear Model: More flexible implementation of the ANOVA technique to test for multiple factors and interactions It does not require equal numbers of data values ('balanced') between each combination of factor levels. See Website
	ANCOVA – Analysis of Covariance: Performs an ANOVA analysis when one variable is also dependent on an additional variable. See Website
	Tukey test: *Post hoc* test to locate the specific differences between samples following a 'significant' ANOVA result. See Example 11.9

Statistic:	Sample variances or standard deviations (two samples)
Test:	*F*-test
	Compares the *variances* of two sets of replicate data Assumes that data is taken from a *normal* distribution. See Example 10.8

Statistic:	Sample medians
Test:	Range of non-parametric tests
	Compares the *median(s)* of sets of replicate data Does *not* make any assumptions about the distribution of the source data
	One-sample Wilcoxon test: Compares the median of the sample data set with a specific value. See Example 12.2
	Two-sample Mann–Whitney *U*-test: Compares the medians of two data sample sets. See Example 12.4
	Paired Wilcoxon test: Compares two sets of replicate data in which pairs of data items from each set have unique relationships. See Example 12.6
	One-way Kruskal–Wallis test: Tests for the effect of *one factor* on the median values of *more than two* sample sets. See Example 12.8
	Two-way Friedman test: Tests for the effects of *two factors* on the median values of *more than two* sample sets. See Example 12.10

Statistic:	'*x* and *y*' values of two variables
Test:	Correlation and regression
	Analysis of data in which one variable is expected to change in response to a change in another variable
	Correlation: Tests whether a significant change in one variable is in proportion to a change in the other variable. See Example 13.2
	Pearson's correlation coefficient assumes that data is taken from a normal distribution
	Spearman's correlation coefficient does *not* make any assumptions about the distribution of the source data

	Regression: Once correlation and causation between two variables is confirmed, a regression analysis will calculate the *magnitude* of the effect that one variable has on the other, and produce a 'best-fit' linear model to describe the interaction. 'Regression' is a *calculation* and not, strictly speaking, a test. See Example 4.8

Statistic:	**Frequencies – counting the numbers in categories**
Test:	**Chi-squared, χ^2, test**
	Tests whether the observed frequencies of events in (more than two) specific categories could have occurred by chance
	Contingency table: Tests whether the distribution of frequencies between categories may depend on another factor. See Example 14.1
	Goodness of fit: Tests whether the distribution of frequencies between categories is consistent with an expected distribution. See Example 14.4

Statistic:	**Proportions**
Test:	**Fisher's exact test**
	Tests whether the observed distribution of frequencies between just two categories could have occurred by chance. Could also be performed by: χ^2-test using the Yates correction – see 14.1.5 Normal distribution approximation – see 14.3.3
	One-proportion: Compares an observed proportion with a specific expected value. See Example 14.6
	Two-proportion: Compares an observed proportion with another observed proportion. See Example 14.10

10

t-tests and *F*-tests

Overview

 • 'How to do it' video answers for all 'Q' questions.
• *t*-tests and *F*-tests in Excel and Minitab.

We saw in 7.2 that the basic characteristics that describe a data set are:

• mean value;
• standard deviation; *or*
• variance = (standard deviation)^2.

In this chapter we introduce the statistical tests that test for differences based on these values:

• The *t*-test looks for differences in mean value (location) between data samples.
• The *F*-test looks for differences in variance (spread) between data samples.

This chapter outlines the interpretation of these tests, using the '*p*-value method', and also introduces the test statistics t_{STAT} and F_{STAT}. It is assumed that readers are already familiar with the general procedures for hypothesis testing that were introduced in 9.3 and 9.4.

In understanding the *t*-test, it is also useful to be familiar with the concept of the confidence interval that was introduced in 8.2.4.

An important assumption with both *t*-tests and *F*-tests is that the underlying populations of data values are normally distributed (see 8.1.3). However, the *t*-test is robust to deviations from normality and can be used reliably with many non-normal distributions, provided that the actual distribution is fairly *symmetrical* and without a long 'tail'.

The 'tests' in this chapter are categorized as 'parametric' because the actual data values are used directly in the calculations. In 'non-parametric' tests (Chapter 12) the data values are used only to *rank* the values in order, and the tests are then based on the ranking of data rather than actual values. It is necessary to use non-parametric tests either for ordinal data or for data that does not follow the normal distribution (see 9.5).

Excel functions and Data Analysis Tools can be used to perform two-sample *t*-tests and *F*-tests – see Appendix I. The Excel functions TINV and FINV can also be used to give values for t_{CRIT} (10.1.3) and F_{CRIT} (10.4.3) respectively.

10.1   One-sample *t*-tests

10.1.1   Introduction

The mean value, $\bar{x}$, of a sample of replicate measurements is the 'best estimate' of the true value (or population mean), μ, of the variable being measured (see 8.2.4). For a very large (ideally infinite) number of replicate measurements the *sample* mean value would become equal to the *true* value, μ.

The one-sample *t*-test compares the mean, $\bar{x}$, of one sample of data with a *specific value*, μ_0, to test whether the true value, μ, of the variable being measured might differ from that specific value, μ_0.

Example 10.1

Four replicate measurements of blood–alcohol level (in mg per 100 mL, see Example 8.1) are made, giving one *statistical* sample with $n = 4$ values:

$$77.7 \qquad 79.9 \qquad 76.8 \qquad 78.0$$

which have a mean value $\bar{x} = 78.1$ and a sample standard deviation $s = 1.3$.

We wish to test whether we could claim with 95 % confidence that the true blood–alcohol level, μ, being measured was actually less than 80 mg per 100 mL.

The analysis is performed in the following text.

The hypotheses for the 1-tailed (see 9.4.3) test outlined in Example 10.1 are as follows.

The proposed hypothesis, H_1, is that the true blood–alcohol, μ, is indeed less than μ_0 (=80), i.e. $\mu < \mu_0$.

The null hypothesis, H_0, is that the observed mean value, $\bar{x}$, actually originated as the mean of n random values from a distribution of values (shown in Figure 10.1) with a true mean of μ_0, i.e. $\mu = \mu_0$.

10.1.2   Using the *p*-value

The use of *p*-values for hypothesis testing has been introduced in 9.3 and 9.4.

If the null hypothesis were true in Example 10.1, then the *possible* mean values for samples of size $n = 4$ would follow a probability distribution as outlined in Figure 10.1.

The shaded area in Figure 10.1 gives the probability that the *experimental* mean value of a sample, $\bar{x}$, might be equal to, or less than, 78.1, even when the true value is given by $\mu_0 = 80$.

Thus the shaded area is the probability, p, of making a Type I error (see 9.4.4), i.e. by deciding that the proposed hypothesis was true, and that the null hypothesis was *not* true. However, when accepting that the proposed hypothesis is true, we must be willing to accept a small probability (significance level, α) that we might make a Type I error.

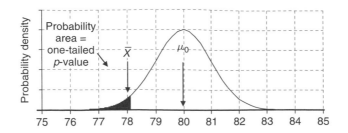

Figure 10.1   Possible variation of sample means.

If the probability, *p*-value, is less than, or equal to, the significance level, $\alpha(p \leqslant \alpha)$, then we accept the proposed hypothesis, that the true value, μ, is not equal to μ_0. The requirement for '95 % confidence' in the question implies a significance level of $\alpha = 0.05$.

The calculation for Example 10.1 is performed on the Website using Minitab and returns a value for the *p*-value equal to 0.031. Since $p < \alpha(0.05)$, we accept the proposed hypothesis and decide that the true blood–alcohol level, μ, is less than 80 mg per 100 mL.

10.1.3   Using the *t*-statistic, t_{STAT}

The t_{STAT} is calculated as the ratio of the *difference* between $\bar{x}$ and μ_0 to the estimated standard uncertainty in the sample mean, $s/\sqrt{n}$ (see equation [8.4]):

$$t_{STAT} = \frac{(\bar{x} - \mu_0)}{s/\sqrt{n}} \qquad [10.1]$$

A *large* difference between $\bar{x}$ and μ_0 would tend to give a *large* value for t_{STAT} (either positive or negative). Hence we will set that criterion that:

H$_1$ is accepted if the positive value of t_{STAT} (ignoring any signs) is *greater* than, or equal to, the critical value, t_{CRIT}, i.e. $|t_{STAT}| \geqslant t_{CRIT}$.

The values for t_{CRIT} are given in Appendix III and depend on:

- whether the test is 1- or 2- tailed;
- the chosen significance level, α (typically 0.05); and
- the degrees of freedom for a one-sample *t*-test, $df = n - 1$.

Using the *t*-statistic method for Example 10.1, and entering the values $\bar{x} = 78.1$, $\mu_0 = 80$, $s = 1.3$ and $n = 3$ into equation [10.1] gives:

$$t_{STAT} = (78.1 - 80.0)/(1.3/\sqrt{4}) = -2.92$$

From Appendix III, for a 1-tailed t-test with $df = 4 - 1 = 3$ and $\alpha = 0.05$ we find that:

$$t_{CRIT} = 2.35$$

Since $|t_{STAT}| \geqslant t_{CRIT}$ we accept the proposed hypothesis, H_1: $\mu < 80.0$.

Q10.1

The mean level of pollutant in the wastewater of an industrial company should not exceed 3.50×10^{-8} g L^{-1}.

An experiment is conducted to decide whether the wastewater has a mean level of pollutant greater than 3.50×10^{-8} g L^{-1}. If it is concluded that it does, then the probability of error in the conclusion should not be greater than 1 in 20.

Replicate analysis of six samples of the water reveal levels of 3.45, 3.65, 3.58, 3.56, 3.68 and 3.59×10^{-8} g L^{-1}.

(i) State the hypotheses appropriate to this problem.
(ii) What is the appropriate level of confidence for this test?
(iii) Calculate the mean and standard deviation of the sample.
(iv) Calculate the t-statistic, t_{STAT}, appropriate to this problem.
(v) Decide if this is a 1-tailed or 2-tailed t-test.
(vi) Calculate the number of degrees of freedom for t_{CRIT}
(vii) Look up the value for t−critical, t_{CRIT}.
(viii) Comment on whether the level of pollutant is greater than 3.50×10^{-8} g L^{-1}.

10.1.4 Applying one-sample t-tests

We will illustrate the use of the one-sample t-test by using the replicate data of the sample sets 13, 14 and 18 from Example 7.2, which are given in Table 7.1 and reproduced here in Table 10.1.

Table 10.1.

Set	Data					Mean, $\bar{x}$	Std dev., s
13	78.0	79.0	81.7	80.7	84.1	80.7	2.384
14	80.8	82.1	80.5	80.6	81.3	81.1	0.658
18	78.6	79.1	76.4	80.0	77.9	78.4	1.355

Example 10.2

Test whether the analysts who record data sets 13, 14 and 18 would conclude (separately) that the true blood–alcohol level, μ, was not equal to (i.e. different from) the specific value, $\mu_0 = 80.0$ mg per 100 mL.

The analysis is performed in the following text.

The relevant 2-tailed hypotheses for the question in Example 10.2 would be

H_1: $\mu \neq 80.0$
H_0: $\mu = 80.0$

We can calculate values for t_{STAT} using equation [10.1], look up values for t_{CRIT} and use software to calculate p-values, leading to the following conclusions:

Analyst	t_{STAT}	t_{CRIT}	p-value	Decision	Conclusion ($\alpha = 0.05$)
13	0.66	2.78	0.547	Do not accept H_1	μ could be 80.0
14	3.60	2.78	0.023	Accept H_1	μ is not 80.0
18	−2.64	2.78	0.058	Do not accept H_1	μ could be 80.0

We know that the 'true' value from which the data samples were drawn was indeed 80.0 (see Example 7.2). Hence we can see that:

- Analysts 13 and 18 have arrived at the correct conclusion.
- However, the data for analyst 14 has, by chance, led to an incorrect Type I error (compare the results of confidence intervals given in Table 7.1 and displayed in Figure 7.4).

Example 10.3

We have seen (in Table 7.1 and displayed in Figure 7.4) that, for data set 18, the confidence interval (with 95 % confidence) is from 76.7 to 80.1. Is this consistent with the conclusion reached by analyst 18 using the 2-tailed t-test that the true value could be 80.0?

The value of 80.0 falls *within* the limits of the calculated confidence interval. Hence it must be accepted that the true value could be 80.0, and this is indeed consistent with the conclusion reached by using the 2-tailed t-test in Example 10.2.

Example 10.4

Analyst 18 then notices that most of the data values in set 18 fall below 80.0 and decides to redo the *t*-test, but this time using a 1-tailed test ($\alpha = 0.05$) to ask whether the true value of the blood–alcohol *is less than* 80.0.

The new, 1-tailed, hypotheses are:

H_1: $\mu < 80.0$
H_0: $\mu = 80.0$

The calculation gives the results:

Analyst	t_{STAT}	t_{CRIT}	p-value	Decision	Conclusion
18	-2.64	2.13	0.029	Accept H_1	μ is less than 80.0

Note that the value of t_{CRIT} has changed with the number of 'tails' implied in the question.

As the *p*-value is now less than 0.05, analyst 18 decides to accept H_1 and concludes that the true blood–alcohol level μ is less than 80.0.

Which of the following is the correct conclusion (they cannot both be right)? Either:

The evidence is sufficient to claim that μ is less than 80.0 (from the 1-tailed test).

or:

There is not enough evidence to claim a difference between μ and 80.0 (from the 2-tailed test in Example 10.2)

The answer is given in the following text.

The 'conclusion' reached in Example 10.4 using the 1-tailed test was invalid, because the analyst used the same data *twice* – once to decide on the direction to test for the 1-tailed test, and then again to actually carry out the test. The choice and direction of any test must be decided independently of the data that is being analysed.

Q10.2

Two new sets of measurements, X and Y, of the same blood–alcohol level as in Example 7.2 give the following values for t_{STAT}, t_{CRIT} and the

p-value. What should be the appropriate decisions and conclusions, assuming $\alpha = 0.05$?

	Set	t_{STAT}	t_{CRIT}	p-value	Decision	Conclusion
	X	−2.05	2.13			
	Y			0.048		

Q10.3

Three scientists perform statistical analyses on the following set of replicate data for the measurement of the true pH of a solution:

$$8.49 \quad 8.46 \quad 8.03 \quad 8.72 \quad 8.96 \quad 9.26 \quad 9.42 \quad 8.17$$

(i) Scientist 1 calculates the 95 % confidence interval correctly as 8.69 ± 0.42. Would scientist 1 claim that the true pH value was not equal to 8.34?
(ii) Scientist 2 performs a 2-tailed t-test on the same data to test whether the true value is not equal to 8.34, and records one of the following p-values:

(a) 0.036   (b) 0.050   (c) 0.088

On the basis of the data in (i), which of the above would be correct?
(iii) Scientist 3 performs a 1-tailed t-test on the same data to test whether the true value is greater than 8.34, and records one of the following p-values:

(a) 0.072   (b) 0.050   (c) 0.044

On the basis of the result in (ii), which of the above would be correct?

10.2 Two-sample t-tests

10.2.1 Introduction

Replicate measurements in two sets, A and B, of sample data will record mean values, $\bar{x}_A$ and $\bar{x}_B$, which are the best estimates of the true means, μ_A and μ_B, respectively, of the populations from which the samples were derived.

The two-sample t-test tests whether the difference between experimental sample means $\bar{x}_A$ and $\bar{x}_B$ is large enough to imply an actual difference between the true values of μ_A and μ_B.

Example 10.5

In a survey of pollution, the chemical oxygen demand (COD) was measured for two sources of wastewater, A and B. Measurements were made on samples from A (six replicates) and B (five replicates). The numerical data values, together with the means and sample standard deviations, are given below:

							Mean	Standard deviation
A	77.8	82.6	80.4	72.4	80.2	76.4	$\bar{x}_A = 78.3$	$s_A = 3.61$
B	73.2	76.8	73.8	71.3	74.8		$\bar{x}_B = 74.0$	$s_B = 2.03$

Assess whether it can be said with 95 % confidence that the COD values for the two sources are different.

The analysis is performed in the following text.

The distribution of the experimental values in the two samples can be represented by boxplots as in Figure 10.2.

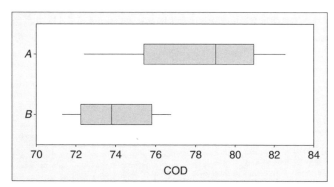

Figure 10.2   Boxplots of COD measurements for Example 10.5 (using Minitab).

In respect of the test outlined in Example 10.5, the 2-tailed hypotheses are:

The proposed hypothesis, H_1, is that the true COD values, μ_A and μ_B, from different sources are different, i.e. $\mu_A \neq \mu_B$.

The null hypothesis, H_0, is that the true COD values are the same (i.e. $\mu_A = \mu_B$), and that the observed mean values, $\bar{x}_A$ and $\bar{x}_B$, actually originated as two random values from a single distribution of values.

If the null hypothesis were true, then *possible* values for the sample means might follow a probability distribution as outlined in Figure 10.3, in which it is possible that the different experimental mean values of $\bar{x}_A$ and $\bar{x}_B$ were recorded even though the true values were the same, $\mu_A = \mu_B$, in both cases.

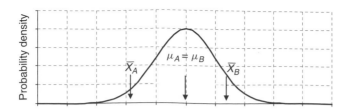

Figure 10.3   Possible variation of sample means.

10.2.2   Using the p-value

The calculation for Example 10.5 is performed on the Website using both Excel and Minitab and returns a value for the *p*-value equal to 0.042. The calculation assessed the probability, *p*, that the observed values, $\bar{x}_A$ and $\bar{x}_B$, might have originated from a single distribution as in Figure 10.3. However, since $p < \alpha(0.05)$, we accept the proposed hypothesis and that the COD values of the two sources, A and B, are actually *different*.

In using a software calculation for a two-sample *t*-test, it is usually necessary to choose whether or not the population standard deviations, σ_A and σ_B, can be assumed to be the same (i.e. equal or unequal variances). Unless there is clear evidence that the spread of one data sample is very different from the other, then it is usual to 'assume equal variances'. If in doubt it is possible to apply an *F*-test (see 10.4) to check for a difference between variances.

10.2.3   Using the t-statistic, t_{STAT}

For a two-sample *t*-test, t_{STAT} is calculated as the ratio of the *difference* between $\bar{x}_A$ and $\bar{x}_B$ to the estimated standard deviation of the possible sample mean values. In this case we assume 'equal variances' for the population sources of the data two samples, and calculate a single pooled standard deviation, s', as an 'average' of the two sample standard deviations, s_A and s_B:

$$\text{Pooled standard deviation, } s' = \sqrt{\frac{(n_A - 1)s_A^2 + (n_B - 1)s_B^2}{(n_A + n_B - 2)}} \qquad [10.2]$$

where n_A and n_B are the numbers of data values in sample A and B respectively.

The t_{STAT} then becomes (compare this with equation [10.1]):

$$t_{STAT} = \frac{(\bar{x}_A - \bar{x}_B)}{s' \times \sqrt{1/n_A + 1/n_B}} \qquad [10.3]$$

A *large* difference between $\bar{x}_A$ and $\bar{x}_B$ would tend to give a *large* value for t_{STAT} (either positive or negative). Hence we will set that criterion that:

H$_1$ is accepted if the positive value of t_{STAT} (ignoring any signs) is *greater* than, or equal to, the critical value, t_{CRIT}, i.e. $|t_{STAT}| \geq t_{CRIT}$.

The values for t_{CRIT} for are given in Appendix III and depend on:

- whether the test is 1- or 2- tailed;
- the chosen significance level, α (typically 0.05); and
- the degrees of freedom for a two-sample t-test, $df = n_A + n_B - 2$.

Using the t-statistic method for Example 10.5, and entering the values $\bar{x}_A = 78.3$, $s_A = 3.61$, $n_A = 6$, $\bar{x}_B = 74.0$, $s_B = 2.03$ and $n_B = 5$:

$$s' = \sqrt{\frac{(n_A - 1)s_A^2 + (n_B - 1)s_B^2}{(n_A + n_B - 2)}} = \sqrt{\frac{(6 - 1) \times 3.61^2 + (5 - 1) \times 2.03^2}{(6 + 5 - 2)}} = 3.01$$

then:

$$t_{STAT} = \frac{(\bar{x}_A - \bar{x}_B)}{s' \times \sqrt{1/n_A + 1/n_B}} = \frac{(78.3 - 74.0)}{3.01 \times \sqrt{1/6 + 1/5}} = 2.37$$

From Appendix III, for a 2-tailed t-test with $df = 6 + 5 - 2 = 9$ and $\alpha = 0.05$ we find that:

$$t_{CRIT} = 2.26$$

Since $|t_{STAT}| \geqslant t_{CRIT}$ we accept the proposed hypothesis, $H_1: \mu_A \neq \mu_B$.

Q10.4

Samples of two powders were analysed for their particle diameters. Use a t-test to test whether the two powders came from original distributions that had the same, or different, mean diameters. The level of confidence should be 95 %.

The results of the measurements (in arbitrary units) were:

Sample A (11 observations): mean = 6.65, standard deviation = 3.91
Sample B (16 observations): mean = 4.28, standard deviation = 2.83

10.2.4 Applying two-sample *t*-tests

We will illustrate the use of the two-sample t-test with Example 10.6.

Example 10.6

The levels of a pollutant in three rivers were analysed giving results as below (each value is multiplied by 10^{-8} g L^{-1} to give the correct magnitudes and units):

River	Data						Mean, $\bar{x}$	Std dev., s
A	6.80	7.25	6.27	6.60	6.37		6.66	0.390
B	7.44	8.52	7.35	8.28	7.15	7.61	7.73	0.549
C	6.64	7.35	7.14	6.55			6.92	0.387

Use 2-tailed *t*-tests to test whether there are any significant differences in the true pollution levels (μ_A, μ_B and μ_C) between each pair of rivers (A, B and C) respectively.

The analysis is performed in the following text.

The distribution of the experimental values in the three samples can be represented by boxplots as in Figure 10.4.

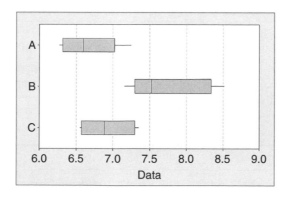

Figure 10.4   Boxplots for pollutant levels for Example 10.6 (using Minitab).

The overall aim of the investigation was to identify any differences in pollution between the rivers.

The relevant 2-tailed hypotheses when comparing rivers A and B (for example) in Example 10.6 would be of the form:

H_1: $\mu_A \neq \mu_B$
H_0: $\mu_A = \mu_B$

We can calculate values for t_{STAT} using equation [10.3], look up values for t_{CRIT} and use software to calculate p-values, leading to the following conclusions:

Comparison pairs	t_{STAT}	t_{CRIT}	p-value	Decision	Conclusion ($\alpha = 0.05$)
A and B	−3.64	2.26	0.005	Accept H_1	$\mu_A \neq \mu_B$
B and C	2.52	2.31	0.036	Accept H_1	$\mu_B \neq \mu_C$
A and C	−1.01	2.36	0.348	Do not accept H_1	No evidence of a difference between μ_A and μ_C

As can be expected from Figure 10.4, there is not enough evidence to suggest a difference between the pollutant levels for rivers A and C. However, the t-tests confirm that there is a significant difference between river B and both rivers A and C, even though there is a small overlap in the ranges of all data samples.

10.3   Paired t-tests

10.3.1   Introduction

A paired t-test is a special form of a two-sample test, in which each data value in one sample can be 'paired' uniquely with one data value in the other sample.

Example 10.7

A student aims to test whether exercise increases systolic blood pressure. Seven subjects measure their systolic blood pressure (BP, in mm Hg) before and after exercise, with the results given in Table 10.2.

Table 10.2.

Subject:	1	2	3	4	5	6	7		Standard
BP before	115	122	145	133	147	118	130	Mean,	deviation,
BP after	125	127	155	136	142	124	139	$\bar{x}_{DIFF}$	s_{DIFF}
Difference	10	5	10	3	−5	6	9	5.43	5.32

The analysis is performed in the following text.

The *differences* in blood pressure are also listed in Table 10.2 for each subject, which indicate that, for all except one subject, the blood pressure does increase.

The paired test effectively tests whether the true mean of the differences (for all subjects), μ_{DIFF}, is greater than zero. This part of the calculation has become a one-sample t-test, comparing the mean of the experimental differences, $\bar{x}_{DIFF}$, with zero.

The relevant 1-tailed hypotheses for the question in Example 10.7 would be:

H_1: Mean differences are greater than zero, $\mu_{\text{DIFF}} > 0$

H_0: Mean differences are zero, $\mu_{\text{DIFF}} = 0$

The degrees of freedom for the paired test are $df = n - 1$, where n is the number of data pairs.

10.3.2 Applying the paired t-test

Treating the differences in Example 10.7 as a one-sample, 1-tailed test, the data values:

$$\bar{x}_{\text{DIFF}} = 5.43, s_{\text{DIFF}} = 5.32, n = 7$$

give the following results:

p-value method: $p = 0.0178$ $p < 0.05$ Accept H_1

t-statistic method: $t_{\text{STAT}} = 2.70, t_{\text{CRIT}} = 1.94$ $t_{\text{STAT}} > t_{\text{CRIT}}$ Accept H_1

Q10.5

If we just enter the same data from Example 10.7 into a two-sample t-test, without taking differences, we obtain a p-value of 0.205.

Which test was the most powerful – the two-sample t-test or the paired t-test?

Q10.6

Seven 'experts' in the taste of real ale have been asked to give a 'taste' score to each of two brands of beer, Old Whallop and Rough Deal. The results, a_i and b_i for each expert are as follows, together with the differences, d_i, between the scores for each expert:

Expert	A	B	C	D	E	F	G
Old Whallop, a_i	60	59	65	53	86	78	56
Rough Deal, b_i	45	62	53	47	65	80	46
Difference, $d_i = a_i - b_i$	15	−3	12	6	21	−2	10

This is a *paired test*, because the identification of the specific expert is the factor that links the values a_i and b_i in the pair of values for each expert.

Assume that the scoring for each expert follows a normal distribution. Complete a paired *t*-test, by carrying out a one-sample *t*-test on the set of seven difference values to test whether the mean of the differences is significantly different (at a level of 0.05) from zero.

The equivalent *non-parametric test*, where the data may not be normally distributed, is given in Q12.3.

10.4 F-tests

10.4.1 Introduction

Sometimes we want to test whether there is a difference in the 'spread', or uncertainty, of two data samples, e.g. testing whether the data from one measurement process shows a significantly greater experimental uncertainty than another.

The direct measure of data uncertainty is given by the *standard deviation* of the data values. However the *F*-test actually compares the *variances* of two data samples, where:

$$\text{Variance} = (\text{standard deviation})^2$$

We see elsewhere (8.3.3 and 11.1.2) that calculations involving experimental uncertainty often use variance.

Example 10.8

Two technicians, A and B, regularly perform the same type of analysis. However, it is suspected that technician B has become less precise in his procedures, leading to a greater variability in the results that he produces. The data in Table 10.3 shows replicate measurements made by each technician on the same material.

Is there evidence to say that the results from B are significantly (at 95 % confidence) more variable than those from A?

Table 10.3.

									Mean	Standard deviation	Variance
A	23.4	24.3	23.4	24.2	23.8	23.8	23.7	24.3	23.86	0.370	0.137
B	23.6	25.4	23.5	25.2	24.3	24.5	24.2	25.3	24.50	0.745	0.554

The analysis is performed in the following text.

The relevant 1-tailed hypotheses for the question in Example 10.8 would be:

H_1: Variance of the B-data is greater than the A-data, $\sigma_B^2 > \sigma_A^2$.
H_0: Variance of the B-data is equal to the A-data, $\sigma_B^2 = \sigma_A^2$.

10.4.2 Using the p-value

In some software (e.g. Minitab) the F-test appears as one option under tests for 'Two Variances'. The calculation for Example 10.8 is performed on the Website using both Excel and Minitab and returns a value for the p-value equal to 0.043. Since $p < \alpha$ (0.05), we accept the proposed hypothesis that the variance of B is greater than that of A.

10.4.3 Using the F-statistic, F_{STAT}

The F-statistic is defined as:

$$F_{STAT} = \frac{s_B^2}{s_A^2} \qquad [10.4]$$

where s_B^2 and s_A^2 are the two *sample* variances being compared.

For a 1-tailed test, the variance that is *expected* to be greater must be placed in the numerator. For a 2-tailed test, the variance that is *seen* to be greater should be placed in the numerator.

H_1 is accepted if the value of F_{STAT} is *greater* than, or equal to, the critical value, F_{CRIT}, i.e. $F_{STAT} \geqslant F_{CRIT}$.

The values for F_{CRIT} are given in Appendix IV and depend on:

- whether the test is 1- or 2- tailed;
- the chosen significance level, α (typically 0.05); and
- the degrees of freedom for the two samples, $df_B = n_B - 1$ and $df_A = n_A - 1$.

Using the F-statistic method for Example 10.8, and entering the values $s_B^2 = 0.554$ and $s_A^2 = 0.137$, gives:

$$F_{STAT} = \frac{0.554}{0.137} = 4.04$$

From Appendix IV, for a 1-tailed F-test with $df_B = 8 - 1 = 7$ and $df_A = 8 - 1 = 7$ and $\alpha = 0.05$ we find that:

$$F_{CRIT} = 3.79$$

Since $F_{STAT} \geqslant F_{CRIT}$ we accept the proposed hypothesis, $H_1 : \sigma_B^2 > \sigma_A^2$.

Q10.7

Find the values for F_{CRIT} for:

(i) A 1-tailed test for a significance level of 0.05 and a numerator with 10 data values and denominator with 8 data values.

(ii) A 2-tailed test for a significance level of 0.05 and a numerator with 16 data values and denominator with 10 data values.

Q10.8

In order to assess whether two examinations, P and Q, had a difference in the spread of their results, it was decided to compare the variance of their results for a sample group of 21 students who took both exams.

Recorded variances were as follows: $s_P^2 = 85.0$ and $s_Q^2 = 195.5$.

Apply:

(i) a 2-tailed test, with a proposed hypothesis which states that there is no difference between the variances of the two exams;

(ii) a 1-tailed test, with a proposed hypothesis which states that the variance of exam P would be less than that of exam Q.

Q10.9

We wish to compare two different methods, X and Y, for measuring the lead content of water. The same sample of water is analysed by each method, giving eight replicate results using method X and seven replicate results using method Y, as given in the table below:

								Mean	Standard deviation	Variance
X	83.3	81.4	81.0	79.8	82.5	84.7	83.2	83.3	82.40	1.57	2.47
Y	81.5	77.9	76.9	74.7	79.9	84.3	81.3		79.50	3.24	10.47

(i) As we can see from the data that $s_Y^2 > s_X^2$, would it be acceptable to use a 1-tailed test when testing for a difference in variance?

(ii) Test for a difference in the *experimental uncertainty* between the two methods.

(iii) Test for a difference in the *mean values* reported by the two methods.

11

ANOVA – Analysis of Variance

Overview

- 'How to do it' video answers for all 'Q' questions.
- Implementation of ANOVAs in Excel and Minitab.
- Additional material: ANCOVAs.

We saw in Chapter 10 how the two-sample t-test tests for a difference between the mean values of *two* data samples.

ANOVAs form a set of analytical tests which can be used to identify possible differences in the mean values of *more than two* data samples.

As its name suggests, an ANOVA uses the analysis of variances in the overall data to identify whether there are factors involved which cause variations beyond the inherent experimental variation in the data values. At its simplest level, the ANOVA performs an F-test to test whether the overall variance in the data values from *all* samples is significantly greater than the inherent experimental variance *within* each sample. If the F-test gives a significant result, then it can be concluded that there is a significant difference *between* the mean values of the samples.

In this chapter we look at some increasing sophistication in the ANOVA, identifying the effects of multiple factors and any interaction between them. The introduction of even more powerful ANOVAs is continued on the Website.

We also discuss what can be done *after* an ANOVA has detected an effect, by introducing the use of *post hoc* tests.

11.1 One-way ANOVA

11.1.1 Introduction

We can illustrate the use of the ANOVA analytical approach by considering the following example.

Essential Mathematics and Statistics for Science 2nd Edition   Graham Currell and Antony Dowman
Copyright © 2009 John Wiley & Sons, Ltd

Example 11.1

A student chooses to compare the growth rates of plants in four different soils, A, B, C and D, and selects seeds at random, planting three in each of the soils. She wishes to test whether there may be a significant difference, at a significance level of $\alpha = 0.05$, between the growth rates in the different soils.

The mean heights measured after a certain time are recorded in the table below:

Soil	Heights			Mean	Std dev.
A	11.9	11.0	11.8	11.6	0.493
B	11.6	11.8	12.2	11.9	0.306
C	9.6	11.1	9.9	10.2	0.794
D	10.4	11.0	10.7	10.7	0.300

The analysis is performed in the following text.

The different heights 'within' each soil sample show a natural variation due to the different growth rates of plants. These can be reproduced graphically as shown in Figure 11.1.

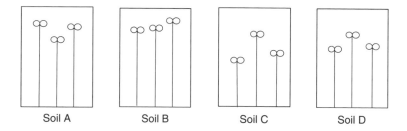

Figure 11.1   Three seedlings in four different soils.

We can also use boxplots to give a visual representation of the data variations (Figure 11.2) within each group.

From the visual representations, there appear to be differences between the four sets of data, in both variability and mean values. As part of an initial guess, we might expect that there is a significant difference between the mean values of B and D because their 'boxes' *do not* overlap, but that there is no significant difference between C and D because their 'boxes' *do* overlap. However, we must use quantifiable tests if we are to make reliable decisions.

The student is familiar with the t-test for testing for differences in mean values and applies a two-sample t-test for the means $\overline{x}_A$ and $\overline{x}_B$ of the two soils A and B. She records a p-value of 0.421, which suggests that there is not a significant difference between the qualities of soils A and B.

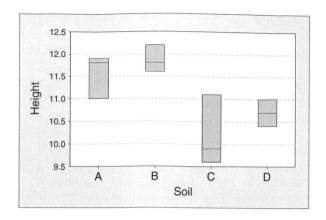

Figure 11.2 Boxplots of the heights of seedlings in four different soils (using Minitab).

As there were four soils under test as shown, each of which has a natural variation, the student performs five further two-sample t-tests between other data pairs, obtaining the *pairwise* results in Table 11.1.

Table 11.1. The p-values for pairwise t-tests between samples.

	B	C	D
A	0.421	0.064	0.060
B		0.027	0.009
C			0.365

These results might suggest that pairs A/B, A/C, A/D and C/D are not dissimilar but that soil B is significantly different from both C and D. However, we will see later (11.4.3) that the use of multiple t-tests in this way can lead to incorrect conclusions. This is because the individual t-tests are only using subsets of the overall data in each calculation. In addition, using multiple t-tests increases the probability that one or more of the tests might give a Type I error.

In the following sections, we will introduce the process of analysis of variance (or ANOVA), which uses a single analytical calculation comparing the mean values of any number of possible samples.

11.1.2 Analysing variances

It can be seen from the boxplots in Figure 11.2 that we can describe the **variation** of the data in two parts:

- The variation **within** each sample (or box). The differences in box widths are due to differences in the random variations within each of the three samples.
- The overall variation **between** samples which includes variation due to the different soils in addition to the natural variation within each soil.

This concept of 'separating' the different variances is the basis on which the analysis of variance technique performs its analysis.

In a simple set of experimental results, as in this example, we can expect to find *two* main sources of variation:

Experimental variation. This variation is inherent in the experimental process (1.2), and will normally be due to:
- *measurement* uncertainty (e.g. variations in instrumental results); or
- *subject* uncertainty due to the natural variation in the subjects being measured (e.g. different growth rates from similar biological systems).

Experimental variations are described by the *experimental variance*: σ_E^2.

Factor variation. This variation may be due to the different factor levels between the samples. In Example 11.1 the 'factor' is the soil and the 'level' is the choice of particular soil. The different 'levels' of the factor are often called different 'treatments'. Factor variations are described by the *factor variance*: σ_F^2.

The variance *within* each sample is *only* dependent on the *experimental uncertainty*, because each measurement is at the same factor 'level'. The variance *within* each sample is therefore an *estimate* for the experimental uncertainty, σ_E^2.

The *total* variance in the observed results, σ_{TOT}^2, is equal to the addition of the separate variances, σ_E^2 and σ_F^2:

$$\sigma_{TOT}^2 = \sigma_E^2 + \sigma_F^2 \qquad [11.1]$$

Note that random variations are combined by adding *variances* (see 8.3.3). Remember that the variance, σ^2, is the *square* of the standard deviation, σ.

Q11.1

The true standard deviation variation in pH between a range of different soil samples is given by $\sigma_F = 0.08$. If the pH meter used to make the measurements has a standard deviation uncertainty of $\sigma_E = 0.05$, estimate the combined standard deviation that might be expected in a large number of repeated measurements from randomly selected soil samples.

11.1.3 Mean squares

We now define two 'mean square' values, equivalent to 'variance', which are normally calculated during an analysis of variance:

- Mean square (within), MS_W, gives a measure of the variance 'within' one level of the factor, and will be dependent *only* on the experimental variation:

$$\text{Mean square(within)}, MS_W = \sigma_E^2 \qquad [11.2]$$

- Mean square (between), MS_B, gives a measure of the variance including *both* experimental and factor effects:

$$\text{Mean square (between)}, MS_B = (n \times \sigma_F^2) + \sigma_E^2 \qquad [11.3]$$

The justification for equation [11.3] is that σ_F^2 is the variance of the true *mean* values for the different sample sets, each of which has n data values. The variance of each *single* value due to the factor variance would then be equal to $n \times \sigma_F^2$ (see central limit theorem, 8.2.3). This variance is then combined with the variance σ_E^2 due to experimental uncertainty to get the overall uncertainty.

11.1.4 Significance of the factor effect

If there were no significant difference effect due to different levels of the factor, then we would expect that σ_F^2 would be zero, and we would find that the variances measured by MS_B and MS_W would be of similar magnitude: $MS_B \approx MS_W$.

On the other hand, if there were a significant factor effect, then MS_B would be significantly greater than MS_W: $MS_B > MS_W$.

We can use an F-test to test that any difference that we might observe has not been caused purely by statistical fluctuations in the experimental data. We use the following hypotheses:

Proposed hypothesis, H_1:   The variance MS_B is greater than MS_W.
Null hypothesis, H_0:   The variance MS_B is equal to MS_W.

The appropriate 1-tailed F-statistic (10.4.3) is:

$$F = \frac{MS_B}{MS_W} \qquad [11.4]$$

The degrees of freedom for this F-test are:

$$\text{Numerator } df_B = k - 1 \text{ and denominator}, df_W = k(n - 1) \qquad [11.5]$$

where k is the number of samples and n the number of replicates in each sample.

The standard ANOVA calculation for Example 11.1 gives the results in the form shown in Table 11.2.

Table 11.2. Excel output from one-way ANOVA calculation.

ANOVA

Source of Variation	SS	df	MS	F	p-value	F crit
Between Groups	5.323333	3	1.774444	6.71714	0.014101	4.066181
Within Groups	2.113333	8	0.264167			
Total	7.436667	11				

Here $MS_B = 1.774$, $MS_W = 0.264$, $k = 4$ and $n = 3$, giving:

$$df_B = k - 1 \Rightarrow 4 - 1 \Rightarrow 3 \text{ and } df_W = k(n-1) \Rightarrow 8$$

The *SS* values are 'sum of squares' and can be calculated by:

$$SS_B = MS_B \times df_B = 1.774\,44 \times 3 = 5.323$$
$$SS_W = MS_W \times df_W = 0.264\,17 \times 8 = 2.113$$

The total 'sum of squares' equals the total *SS* values of the 'Between' and the 'Within' groups. It can be seen that the *F*-value is given by the ratio:

$$F = \frac{1.7744}{0.264\,17} \Rightarrow 6.717$$

The 1-tailed value for $F_{CRIT} = 4.07$ can be found from Appendix IV for degrees of freedom $df_B = 3$ and $df_W = 8$, at a significance of $\alpha = 0.05$. The value is also given in the table of results.

Since $F > F_{CRIT}$ we accept the proposed hypothesis that the overall variance in the results is greater than the experimental variance alone.

We can come to the same conclusion, and accept H_1, by noting that the p–value, 0.014, is less than the significance level, $\alpha = 0.05$.

Having accepted that the overall variance in the results, MS_B, IS greater than the experimental variance, MS_W, on its own, the further conclusion is that **there is a significant difference between some (or all) of the sample mean values**.

The ANOVA has detected a difference between some (or all) of the samples, but at the moment we do not know where these difference(s) might lie. The next step will be to apply a *post hoc* test to find out which sample might be significantly different from other samples–see 11.4.2.

Example 11.2

An example is given in the table below, showing the efficiency of a chemical process using three different catalysts (A, B and C) on each of four days:

Catalyst	Day 1	Day 2	Day 3	Day 4
A	84	83	79	80
B	78	79	78	76
C	83	80	78	78

A one-way ANOVA performed in Excel (with the data organized in rows) gives the following results:

ANOVA						
Source of Variation	*SS*	*df*	*MS*	*F*	*p-value*	*F crit*
Between Groups	28.16667	2	14.08333	3.292208	0.084525	4.256492
Within Groups	38.5	9	4.277778			
Total	66.66667	11				

The analysis is performed in the following text.

In Example 11.2, the 'Groups' refer to the different catalysts. The source of variation of the 'Within Groups' is dependent only on the experimental variance, but the source of variation of the 'Between Groups' depends on variances due to the different catalysts, *in addition to* the experimental variance.

From the results, we can conclude that the one-way analysis of this data is *unable* to detect any significant 'catalyst' effect, because $p > 0.05$ and $F < F_{CRIT}$.

Q11.2

In an ecological study of the lifespan of a certain species of protozoan, 50 individuals were divided into five groups of 10. Each group was fed on a different feed. The results of a one-way ANOVA at a significance level of $\alpha = 0.05$ are shown below with some of the data removed:

ANOVA						
Source of Variation	*SS*	*df*	*MS*	*F*	*p-value*	*F crit*
Between Groups	67.7072					
Within Groups	236.4800					
Total	304.1872					

    (i) Calculate the missing values for df_B, df_W, MS_B, MS_W, F, F crit

    (ii) On the basis of your results in (i) decide whether the missing p-value would be greater than, or less than, 0.05.

11.2   Two-way ANOVA

11.2.1   Introduction

We have already seen how a simple one-way ANOVA can identify differences in the mean values or several samples. We now see how the ANOVA technique can begin to unravel more complex problems.

Example 11.3

Two students, Andrew and Bernard, measured the time in minutes that it took them to solve similar sets of statistics problems in the morning and then in the afternoon. The results, together with mean times for rows and columns, are given below:

	Morning	Afternoon	Mean time
Andrew	11.0	16.0	*13.5*
Bernard	15.5	20.0	*17.75*
Mean time	*13.25*	*18.0*	

Does the data show any significant factor differences?

The analysis is performed in the following text.

At first glance it appears that Andrew takes less time than Bernard. However, if we use a two-sample, 2-tailed t-test to test for a difference in the 'student' factor between Andrew

(11.0, 16.0) and Bernard (15.5, 20.0) we find $p = 0.334$, suggesting that there is no significant difference between the students. Using a one-way ANOVA on the same data we obtain the results in Table 11.3, giving the same p-value of 0.334.

Table 11.3. One-way ANOVA for 'Student' factor.

Source of Variation	SS	Df	MS	F	p-value	F crit
Between Groups	18.0625	1	18.0625	1.596685	0.333717	18.51282
Within Groups	22.625	2	11.3125			
Total	40.6875	3				

Similarly, it appears that, on average, the students take less time in the morning than in the afternoon. However, using either a t-test or one-way ANOVA to test for the 'time of day' factor between Morning (11.0, 15.5) and Afternoon (16.0, 20.0) we find $p = 0.255$, which suggests that again there is no significant difference.

We can get a visual impression of the data by plotting the results on a two-dimensional 'interaction' plot in Figure 11.3.

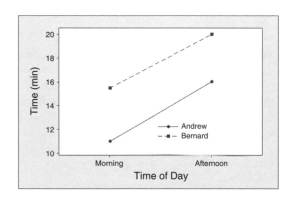

Figure 11.3 Plot of times taken to solve statistics problems (using Minitab).

Looking at the data in just one dimension (i.e. only along the performance time axis), we might see why the single factor tests fail to find any significant differences:

- If we look for a 'student' effect, the differences in times between morning and afternoon create large data variations which reduce the chance of identifying the effect due to the different student.
- Similarly, testing for an effect due to the 'times of day' factor is masked by the variations between 'students'.

11.2.2 Two-way ANOVA

The problem of one factor masking the effect of a second factor can be overcome by using a single *two*-way ANOVA which tests *simultaneously* for the possible effects of *two* factors. Applying a two-way ANOVA to Example 11.3 produces the output given in Table 11.4.

Table 11.4. Two-way ANOVA for 'Student' and 'Time of Day' factors.

Source of Variation	SS	Df	MS	F	p-value	F crit
Student	18.0625	1	18.0625	289	0.037405	161.4462
Time of Day	22.5625	1	22.5625	361	0.033475	161.4462
Error	0.0625	1	0.0625			
Total	40.6875	3				

The two-way ANOVA has analysed the data on the basis of two possible *factors*, each of which has two *levels*:

Factor	Levels
Student	Andrew/Bernard
Time of Day	Morning/Afternoon

The output from the two-way ANOVA looks similar to the one-way ANOVA (Table 11.3), but with the addition of a second factor. The *MS* variance values are calculated for both factors. The remaining variance, MS_{ERROR}, includes the inherent uncertainties in the data, plus an additional variance due to interactions between the factors (see 11.3).

As with the one-way ANOVA, the significance of each factor is calculated by performing an *F*-test to compare each factor variance independently with MS_{ERROR}.

The two-way ANOVA gives *p*-values:

$p = 0.037$ for the 'Student' factor; and
$p = 0.033$ for the 'Time of Day' factor.

Hence both factors are now found to have a significant effect (with $\alpha = 0.05$) on the time it takes to solve the problems.

The two-way ANOVA has taken both factors into account *simultaneously* and can identify the 'two-dimensional' difference that occurs when plotting the data against both factors:

• The lines in Figure 11.3 representing each student are separated, showing a difference in performance times.
• Both lines also show an increase in times between morning and afternoon.

11.2.3 Applying the two-way ANOVA (without replication)

The data in Example 11.3 has no *replication*, i.e. there is only a single experimental measurement for each combination of conditions (of 'Student' and 'Time of Day'). We will see in 11.3 how repeated (replicate) measurements can provide another level of information in the results of the analysis.

Example 11.4

This example uses the same data as Example 11.2, showing the efficiency of a chemical process using three different catalysts (A, B and C) on each of four days:

Catalyst	Day 1	Day 2	Day 3	Day 4
A	84	83	79	80
B	78	79	78	76
C	83	80	78	78

The analysis using a two-way ANOVA (without replication) is performed in the following text.

In the one-way ANOVA analysis of the data in Example 11.2, it was assumed that there is no difference between the conditions, from day to day, under which the *replicate* measurements were made. It was assumed that within each catalyst sample the only variation was the experimental variation, with the result that no significant effect due to the catalyst was found.

However, by using the 'interactions' plot in Figure 11.4, a visual examination of the data suggests that the values tend to decrease from day 1 to day 4. Is this likely to be an actual outcome of randomized data, or could there be some effect due to the 'day' in our experiment that was not taken into account? For example, we may suspect that one of the reagents in the experiment may be deteriorating over the four days and affecting the efficiency of the process.

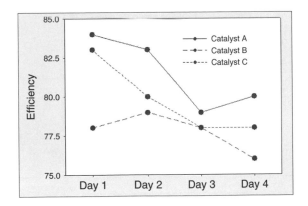

Figure 11.4 Plot of data from Example 11.4 (using Minitab).

We can introduce the 'day' as a second factor in the analysis, and each column would become a different 'level' in that second factor. In terms of experiment design we are now *blocking* these 'replicates' in an organized way, with each block representing different days. The experimental design that *blocks* the *replicate* samples is called a randomized block design – see 15.1.4.

The results from a two-way ANOVA analysis are given in Table 11.5.

Table 11.5.   Excel results for the two-way ANOVA for Example 11.4.

Source of Variation	SS	df	MS	F	p-value	F crit
Catalyst	28.16667	2	14.08333	8.59322	0.017328	5.143249
Day	28.66667	3	9.555556	5.830508	0.032756	4.757055
Error	9.833333	6	1.638889			
Total	66.66667	11				

The results now indicate that there is a significant 'catalyst' effect, because $p = 0.017$, giving $p < 0.05$. The results also show a significant variation due to the 'day', because $p = 0.033$, giving $p < 0.05$.

By comparing the results in Example 11.2 with those in Example 11.4, the one-way ANOVA was unable to detect a significant catalyst effect due to the experimental uncertainty introduced by the variation between days. The two-way ANOVA is now able to identify the contribution by the 'blocks' (or days) to the total variance, and can therefore perform a more sensitive test for the significance of the variance due to the catalyst.

Q11.3

As part of a project in forensic science, a student measures the times to emergence for larvae collected from a recently deceased pig. The times given in the table below were recorded for larvae collected on different days (Day1, Day2 and Day3) and grown on different media (M1, M2, M3 and M4).

	Day1	Day2	Day3
M1	27	19	33
M2	25	24	32
M3	20	19	24
M4	21	20	28

Analyse the data to investigate whether 'day of collection' and/or 'media' have any significant effect (at 0.05) on the emergence times of the larvae.

11.3   Two-way ANOVA with Replication

11.3.1   Introduction

In Example 11.3, we were able to see how a two-way ANOVA was able to detect significant effects due to two separate factors. We now investigate, with Example 11.5, a more complex situation where the effect of one factor depends on the level of the other factor.

Example 11.5

This example has very similar data to Example 11.3. Andrew has the same results as before, but Bernard is replaced by his twin sister, Carol. Carol is quicker at statistics in the afternoon than in the morning, as shown below:

	Morning	Afternoon	Mean time
Andrew	11.0	16.0	*13.5*
Carol	20.0	15.5	*17.75*
Mean time	*15.5*	*15.75*	

Does the data show any significant factor differences?

The analysis is performed in the following text.

A two-way ANOVA analysis of the data in Example 11.5 gives the output shown in Table 11.6.

Table 11.6. Two-way ANOVA for 'Student' and 'Time of Day' factors.

Source of Variation	SS	df	MS	F	p-value	F crit
Student	18.0625	1	18.0625	0.800554	0.535331	161.4476
Time of Day	0.0625	1	0.0625	0.00277	0.966525	161.4476
Error	22.5625	1	22.5625			
Total	40.6875	3				

The results in Table 11.6 show no significant effects for either the 'Student' or 'Time of Day' factors. The interaction plot in Figure 11.5 might give us a clue to understanding the data:

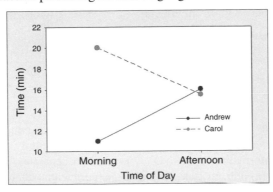

Figure 11.5 Plot of times taken to solve problems (using Minitab).

The plot in Figure 11.5 appears to show that, 'on average', Andrew takes less time than Carol, but, while Andrew takes more time in the afternoon than in the morning, Carol's performance is reversed.

We can see that the effect of the 'Time of Day' factor depends on the 'Student' factor. There *appears to be* an *interaction* between the two factors. However, with the data as given, the two-way ANOVA calculation is unable to separate the data variation due to the possible *interaction* from the variations due to random experimental uncertainty – both are included in MS_{ERROR}.

We will see in the next section that the analytical process can be made more powerful by collecting additional data. It is possible to separate the effect of the interaction from the experimental uncertainty by estimating the magnitude of experimental uncertainty through *repeated data values*, or *replicates*, for measurement.

11.3.2 Two-way ANOVA with replication

In the context of an ANOVA, a *replicate* measurement requires repeating a measurement under exactly the same combinations of conditions. For example, in Example 11.5 this would require at least two measurements for *each* combination of student and time of day.

Example 11.6

The data in the table below has the same mean values as Example 11.5, but now includes two replicate measurements for each factor combination:

	Morning	Afternoon	Mean time
Andrew	11.3, 10,7	15.8, 16.2	13.5
Carol	21.2, 18.8	15.2, 15.8	17.75
Mean time	*15.5*	*15.75*	

Does the data show any significant effects?

The analysis is performed in the following text:

We can perform a two-way ANOVA *with replication*, to obtain the output in Table 11.7.

The replicate data has enabled the ANOVA to estimate the average variance *within* each factor combination, giving an estimate of the inherent experimental uncertainty in each measurement. With this information the ANOVA is able to separate the variance due to the interaction, $MS_{INTERACTION}$, from that due to the experimental uncertainty, which is given by the mean square (within) value, MS_W. Without replication data, these two variances had been combined in Table 11.6 in the MS_{ERROR} value.

Table 11.7. Two-way ANOVA for Example 11.6.

Source of Variation	SS	df	MS	F	p-value	F crit
Student	36.125	1	36.125	43.5241	0.002735	7.708647
Time of Day	0.125	1	0.125	0.150602	0.717729	7.708647
Interaction	45.125	1	45.125	54.36747	0.001803	7.708647
Within	3.32	4	0.83			
Total	84.695	7				

The results in Table 11.7 are:

$p = 0.0027$ for the 'Student' factor;

$p = 0.7177$ for the 'Time of Day' factor; and

$p = 0.0018$ for the interaction between the two factors.

Hence there is a significant difference (with $\alpha = 0.05$) between the two students, and a strong interaction between the performance of the student and the time of day. On average, between the two students, the time of day is not a significant factor.

Q11.4

The percentage yields from a chemical reaction are recorded as functions of both temperature and pressure. Three replicate measurements were made at each of three different pressures, $P1$, $P2$ and $P3$, and two different temperatures, $T1$ and $T2$, as recorded in the table below:

Temperature	Pressures		
	$P1$	$P2$	$P3$
$T1$	79	68	62
$T1$	76	71	65
$T1$	77	69	65
$T2$	76	69	68
$T2$	72	70	71
$T2$	73	73	69

Analyse the above data, using a two-way ANOVA with replication, to assess whether temperature, pressure and/or an interaction between the two have significant effects on the efficiency of the process.

11.3.3   Applying the two-way ANOVA with replication

Example 11.7

The table below gives the relative efficiency of two different motor engines when run on three different varieties of fuel, A, B, C. There are three *replicate* measurements for each combination of engine and fuel.

Fuel mixture

	A	B	C
	66.2	67.2	73.1
Engine 1	65.3	68.4	71.8
	68.0	69.6	70.9
	78.8	78.0	76.6
Engine 2	76.1	76.0	77.6
	79.8	77.0	75.4

Analyse the data to investigate for significant effects on efficiency due to:

(i)   type of engine used;
(ii)   variety of fuel used;
(iii)   interaction between the variety of fuel and the type of engine.

In Example 11.7, the three replicate measurements give additional information (compared with the basic two-way ANOVA) that enables the *experimental* variation to be calculated directly. The ANOVA can then calculate the effect of any possible *interaction* between the type of engine and fuel used.

A two-way ANOVA with replication gives the results in Table 11.8 for Example 11.7.

Table 11.8.   Two-way ANOVA for Example 11.7.

Source of Variation	SS	df	MS	F	p-value	F crit
Engine	310.8356	1	310.8356	178.8124	1.43E-08	4.747221
Fuel	11.89333	2	5.946667	3.420901	0.066735	3.88529
Interaction	38.35111	2	19.17556	11.031	0.001912	3.88529
Within	20.86	12	1.738333			
Total	381.94	17				

The conclusions that can be drawn from Table 11.8 are that there is:

- Significant difference between the performance of the two engines, since $p = 1.43 \times 10^{-8}$ < 0.05. Excel also gives the overall averages (given in Example 11.8) for the engine efficiencies, which show that Engine 2 is more efficient than Engine 1.
- Significant *interaction* between the engine and the fuel, since $p = 0.0019 < 0.05$. The calculated averages in Example 11.8 show that Fuel C *increases* the efficiency of Engine 1 but *decreases* the efficiency of Engine 2 – this is an interaction between engine and fuel.
- No significant *overall* difference between the effects of the fuels, since $p = 0.0667 > 0.05$.

It is possible to represent the interaction between two factors graphically using an interaction plot. In this plot, the mean outcome for each level of one factor is plotted against the levels of the other factor.

Example 11.8

Taking the data from Example 11.7, the mean values are as given in the table below:

	A	B	C
Engine 1	66.5	68.4	71.93
Engine 2	78.23	77	76.53

For each engine a 'line' is drawn on an Excel chart to give efficiency as a function of fuel. The two lines (Figure 11.6) represent the two different engines.

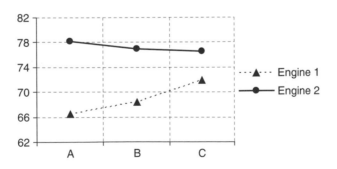

Figure 11.6   Interaction plot for engine efficiences.

From the graph in Figure 11.6 it can be seen that:

- Engine 2 is more efficient than Engine 1 for all fuels, which agrees with $p < 0.05$ for the 'Engine' row in Table 11.8.

- There is an interaction between the fuels and engine – changing from Fuel A to Fuel C improves the performance for Engine 1, but degrades the performance for Engine 2, which agrees with $p < 0.05$ for the 'Interaction' row in Table 11.8.
- The effect of changing the fuel has no 'overall' effect as the effects are opposite for the two engines, which agrees with $p > 0.05$ for the 'Fuel' row in Table 11.8.

Interaction plots can be interpreted by identifying the following:

- A significant effect due to the factor defined by the different 'lines' will give differences between the *vertical positions* of the lines. For example, this is the case with the significant difference in performance between the two engines in Figure 11.6.
- A significant interaction effect will cause the *slopes* of the lines to be different. This is the case with the interaction between engine and fuel.
- A significant effect due to the abscissa (x) variable will give an *overall slope up or down*. In this case, one line goes up and the other down, resulting in no significant overall fuel effect.

The interaction plot can also be redrawn with the other factor as the abscissa.

Q11.5

 Redraw the interaction plot in Example 11.8 using the 'engine' as the abscissa variable.

11.4   ANOVA *Post Hoc* Testing

11.4.1   Introduction

We introduced the ANOVA as a single analytical procedure that detected whether there was a significant factor effect present when comparing the means of several data samples. However, the ANOVA itself does not then identify *which* sample(s) might be different from the others.

Example 11.9

A one-way ANOVA applied to the four soil samples A, B, C and D in Example 11.1 shows that there is a difference between the seedling heights in some (or all) of the soils.

Investigate where the significant difference(s) exist, i.e. which soil is significantly different from which other soil?

The analysis is performed in the following text.

In Example 11.9, we find that comparative boxplots, such as Figure 11.2, can be useful in giving an initial indication of the possible differences. However, it is necessary to use *quantitative* methods to decide whether apparent differences can be considered as significant.

The procedures for locating significant differences are called *post hoc* testing, and can be achieved through a range of possible quantitative *post hoc* tests. In this unit we will introduce the Tukey *post hoc* test, but other options include the Scheffé, Sidak, Dunnett tests, etc.

11.4.2 Tukey *post hoc* test

The Tukey test compares the mean values, $\bar{x}_1$ and $\bar{x}_2$, of each possible pair of samples in turn – a *pairwise* comparison. In each case it applies a test with the hypotheses:

Proposed hypothesis, H_1:   Means are different, $\bar{x}_1 \neq \bar{x}_2$.
Null hypothesis, H_0:   Means are not different, $\bar{x}_1 = \bar{x}_2$.

The option to include a Tukey comparison (e.g. in Minitab) when performing an ANOVA analysis of Example 11.1 results in the *p*-values given in Table 11.9 for possible differences between each pair of soils:

Table 11.9. The *p*-values for pairwise Tukey tests between samples.

	B	C	D
A	0.889	0.046	0.243
B		0.017	0.091
C			0.648

These results suggest that, at a significance of $\alpha = 0.05$, differences exist between the means of pairs A/C and B/C.

11.4.3 Comparison with multiple *t*-tests

In 11.1.1, we performed pairwise *t*-tests between the soil samples in Example 11.1 and obtained the *p*-values in Table 11.10.

Table 11.10. The *p*-values for pairwise *t*-tests between samples.

	B	C	D
A	0.421	0.064	0.060
B		0.027	0.009
C			0.365

The multiple *t*-test approach suggests that significant differences lay between pairs B/C and B/D, whereas the Tukey test puts the differences between pairs A/C and B/C.

The Tukey test is the most reliable because, for each comparison, it calculates the experimental uncertainty from *all* of the available data. The *t*-tests only use the data available from the specific pair of samples used.

For example, the samples B and D both happen to have narrow ranges of data values (Figure 11.2), which will imply that the random experimental uncertainties are less than they actually are, leading to a false significance for the difference between pair B/D. Similarly the *correct* (Tukey) difference between pair A/C is masked in the *t*-test by the wider data spread in both samples A and C.

Q11.6

The yield of a chemical reaction is measured at four different temperatures, with three replicate measurements at each temperature, as in the table below:

	T1	T2	T3	T4
	67.2	66.4	62.3	63.2
	62.4	63.2	59.1	62.4
	66.4	64	60.7	62.4
Means =	65.33	64.53	60.70	62.67

Use an ANOVA with Tukey comparison to identify if the reaction yields at any particular temperatures are significantly different (at 0.05) from other temperatures.

12

Non-parametric Tests for Medians

Overview

- 'How to do it' video answers for all 'Q' questions.
- Wilcoxon, Mann–Whitney, Kruskal–Wallis and Friedman tests in Minitab.

Chapter 9 introduced the concept of a 'hypothesis test', and Chapters 10 and 11 developed the *parametric* *t*-tests and ANOVAs for performing hypothesis tests for differences in mean values. The term **parametric** refers to the fact that the data values are used directly in the calculations for the various test statistics. This can be contrasted with **non-parametric** tests where the numerical values are only significant in establishing the 'ranking', or order, of the data values.

In this chapter, we introduce a range of *non-parametric* tests for differences in median values: the Wilcoxon, Mann–Witney, Kruskal–Wallis and Friedman tests. The Wilcoxon tests are the non-parametric equivalents of one-sample *t*-tests and paired *t*-tests, and the Mann–Whitney test is equivalent to a two-sample *t*-test. The Kruskal–Wallis and Friedman tests analyse experimental data with more than two samples and are the non-parametric equivalents of the one-way and two-way ANOVAs (Chapter 11).

Non-parametric statistics *must* be used for *ordinal data* (Chapter 2 Overview), which does not have an inherent numerical value, but does have a sense of progression. For example, 'opinion' scores in a questionnaire may be ranked 1 = excellent, 2 = good, 3 = satisfactory, 4 = poor, 5 = bad, but these values cannot be used directly in calculations of parametric variables such as *mean* and *standard deviation*. Non-parametric statistics use the *median* value and *interquartile range* (7.1.2) as the equivalent measures of location and spread.

Non-parametric tests can sometimes be used in place of their parametric equivalents for *quantitative* data. For example, the *parametric* data values, 21.2 m, 23.7 m and 22.3 m can be *ranked* in order as 1, 3 and 2 respectively, and the data could be used in either parametric or non-parametric tests. However, due to the loss of information in transforming parametric data to non-parametric data, the non-parametric tests are usually *less powerful* (9.4.5) then their parametric equivalents.

The only data type (Chapter 2 Overview) that *cannot* be ranked is nominal data, because it has no sense of progression from one category to the next, e.g. categorizing people on the basis of their preferences for different sports.

Before proceeding with the following tests, it is recommended that students confirm that they have a good understanding of the principles of hypothesis testing developed in 9.3 and 9.4.

The standard version of Excel does not have the functions or tools to perform non-parametric tests directly. However, dedicated statistics software can perform non-parametric tests easily, albeit with a variety of approaches between the different packages.

The answers, on the Website, to the examples given in the chapter demonstrate the indirect use of Excel in performing non-parametric calculations, in addition to the direct use of statistics software.

Example 12.1

Eight science students (a to h) were asked to what extent they agreed with the following three statements:

S1: The skills for rearranging equations were easy to learn.
S2: The theory of statistics was easy to understand.
S3: The techniques of data analysis in Excel were easy to master.

They were required to mark their answers on a Likert scale from -2 to $+2$:

Disagree strongly	Neutral			Agree strongly
-2	-1	0	$+1$	$+2$

Their results are given in Table 12.1.

Table 12.1.

	Statements		
Student	S1	S2	S3
a	2	0	1
b	0	-1	0
c	1	-1	2
d	2	0	2
e	0	-2	1
f	-1	-1	1
g	-2	-2	-1
h	0	-1	1

This example has been developed for illustrative purposes only. The sample of only eight students would be too small to provide robust results, but are sufficient here to illustrate the use of the relevant statistics:

A visual representation of the data is given in the text.

Figure 12.1 shows the variation between students, with some giving generally higher scores than other students. We can also see that all students give their lowest agreement with statement S2, and, except for student 'a', they give highest agreement with statement S3. The relative score for statement S1 varies from student to student.

Figure 12.2 shows that, *overall*, S2 got the lowest score and S3 the highest.

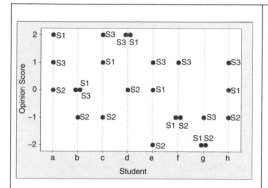

Figure 12.1 Responses to the three statements for each student (using Minitab).

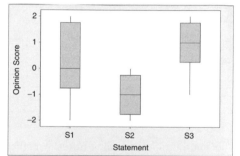

Figure 12.2 Boxplots of the responses for each of the three statements (using Minitab).

Later in this chapter, we will use this example to investigate the use of:

- the one-sample Wilcoxon test to find whether any statement gives a median value significantly different from the neutral response of '0';
- the two-sample Mann–Whitney test to find significant differences in the median values of pairs of student responses;
- the paired Wilcoxon test compared with the two-sample Mann–Whitney test;
- the Kruskal–Wallis test for differences in the medians of the three or more statements; and
- the Friedman test for differences between students, by blocking responses to different statements.

12.1 One-sample Wilcoxon Test

12.1.1 Introduction

The Wilcoxon test is a **one-sample** test – the non-parametric equivalent of the one-sample *t*-test (10.1). The Wilcoxon test can also be used within a *paired* test in the same way that a one-sample t-test can be used within a paired t-test (10.3). Due to the strong parallels between the Wilcoxon test and the one-sample t-test, it would be useful for the readers first to review the use of t-tests.

12.1.2   Test statistic, W

The one-sample Wilcoxon test aims to test whether the observed data sample has been drawn from a population that has a median value, m, that is significantly different from (or greater/less than) a *specific value*, m_0. The tests may be 1-tailed or 2-tailed.

A one-sample Wilcoxon test compares the median of the sample data with a specific value, to test whether the true value (or population median), m, of the variable being measured might differ from that specific value.

The procedure for this test is illustrated using Example 12.2.

Example 12.2

The generation times (5.2.4), t, of 10 cultures of the same micro-organisms were recorded.

Time, t(h)	6.3	4.8	7.2	5.0	6.3	4.2	8.9	4.4	5.6	9.3

The microbiologist wishes to test whether the generation time for this micro-organism is significantly greater than a *specific value* of 5.0 hours.

The analysis is performed in the following text.

We can display the data in Example 12.2 using a box and whisker plot (Figure 12.3).

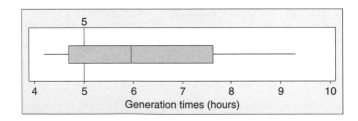

Figure 12.3   Boxplot of data in Example 12.2 (using Minitab).

The test required in Example 12.2 is 1-tailed, i.e. testing for '*greater than* 5.0'. The hypotheses for the Wilcoxon test are therefore:

Proposed hypothesis, H_1: $m > 5.0$

Null hypothesis, H_0: $m = 5.0$

We choose the significance level $\alpha = 0.05$.

The next step is to take the *differences* $(t_i - m_0)$ between each data value, t_i, and the target median, $m_0 = 5.0$. The differences are then *ranked in order of their absolute values* (i.e. ignoring the plus or minus signs) giving the results in Table 12.2. The sign of each difference is retained in the last row.

Table 12.2. Ranking of data for Example 12.2.

Time, t(h)	6.3	4.8	7.2	5.0	6.3	4.2	8.9	4.4	5.6	9.3
Differences $(t - m_0)$	1.3	−0.2	2.2	0.0	1.3	−0.8	3.9	−0.6	0.6	4.3
Ranks (ignoring sign)	5.5	1	7		5.5	4	8	2.5	2.5	9
Sign	+	−	+		+	−	+	−	+	+

Three main **rules** should be observed when calculating rank values:

1. Data items that have a **zero difference** should be excluded from the ranking process and ignored in subsequent calculations. For example, in Table 12.2, one item has a zero difference and is then ignored, leaving only nine data values for subsequent calculations.
2. Data items are ranked in order of their **absolute** values (i.e. ignoring their signs).
3. If two (or more) data items have the **same absolute values (ties)**, then the rank value is *shared* between the values. For example in Table 12.2, the differences '−0.6' and '+0.6' are the joint second and third values and receive the *shared* ranking of 2.5. Similarly the difference '1.3' occurs twice and the two values are given equal ranking of 5.5.

The **test statistic**, $W(+)$, is the *sum* of the rankings for all *positive* differences.
Similarly, the **test statistic**, $W(-)$, is the *sum* of the rankings for all *negative* differences.
The test statistics can be calculated from Table 12.2:

$$W(+) = 5.5 + 7 + 5.5 + 8 + 2.5 + 9 = 37.5$$

$$W(-) = 1 + 4 + 2.5 = 7.5$$

It should be noted that the sum of all rank values is given by:

$$W(+) + W(-) = 0.5 \times n \times (n+1) \qquad [12.1]$$

where n is the number of values included in the calculation, *after any with zero differences have been excluded*.
In Example 12.2, $n = 9$, giving from (12.1):

$$W(+) + W(-) = 0.5 \times n \times (n+1) = 45$$

which is in agreement with the sum of the calculated values.

If the sample median, m, is close to the specific value, m_0 (null hypothesis), we would expect the values of $W(+)$ and $W(-)$ to be very similar.

However, for the proposed hypothesis to be true, we would expect there to be a significant difference between $W(+)$ and $W(-)$, with *one of them* having a significantly *low value*.

The critical values normally given for the Wilcoxon test are the lower critical values, W_L. Appendix V gives critical values at $\alpha = 0.5$ for sample sizes, n.

The conditions for accepting the proposed hypothesis, H_1, are:

$$\text{Accept } H_1 \text{ if} \quad \text{either } W(+) \leqslant W_L \text{ or } W(-) \leqslant W_L.$$

In Example 12.2, the lower critical value, W_L, for a 1-tailed test with $n = 9$ and $\alpha = 0.5$ is $W_L = 8$. Since $W(-) < W_L$ we accept the proposed hypothesis, H_1.

A 1-tailed Wilcoxon test for the problem in Example 12.2 is performed on the Website using Minitab and returns a p-value of 0.043. Since $p < 0.05$, we again accept the proposed hypothesis, H_1.

12.1.3 Applying the one-sample Wilcoxon test

For problems involving ordinal data, the data is already ranked.

Example 12.3

Referring to the data in Example 12.1 presented in Chapter 12 Overview, we now wish to test separately whether the medians, m, for the student responses for each statement, S1, S2 and S3, in Table 12.1 are significantly different from the specific value, $m_0 = 0$.

The analysis is performed in the following text.

The relevant 2-tailed hypotheses for the question in Example 12.3 would be:

$$H_1: \quad m \neq 0$$

$$H_0: \quad m = 0$$

Using Minitab on the Website to perform a 2-tailed Wilcoxon test we obtain the values in Table 12.3 for $W(+)$, $W(-)$, W_L and p-value.

When the p-value results in Table 12.3 are compared with the boxplots in Figure 12.2:

- The conclusions for statements S1 (m could be 0) and S2 ($m \neq 0$) agree with the visual boxplot display.
- It would be difficult to make a decision from the boxplot data for S3 as the values overlap the value $m_0 = 0$. The p-value is close to 0.05, but there is just not enough evidence to accept H_1.

Table 12.3. Results from Example 12.3.

Statement	$W(+)$	$W(-)$	W_L	p-value	Decision	Conclusion
S1	9.5	5.5	$-$ *	0.686	Not accept H_1	True median, m, could be 0
S2	21	0	0	0.036	Accept H_1	$m \neq 0.0$
S3	39	3.0	2	0.076	Not accept H_1	True median, m, could be 0

*After taking tied values, the sample size for S1 is too small for a W_L-value.

Q12.1

In a survey to assess whether 18 trainees found a particular exercise regime useful, they were asked to reply on a scale from -5 (not at all useful) to $+5$ (very useful). Their scores are listed below:

3	5	-3	5	4	-2	2	4	-1
0	5	0	-4	-2	3	0	1	3

(i) Use a non-parametric test to assess whether the results show that the regime was considered to be useful (i.e. the median score is greater than '0').

(ii) What result would have been obtained for a test to assess whether the results show a median value which is not equal to '0', i.e. either greater than or less than '0'?

12.2   Two-sample Mann–Whitney *U*-test

12.2.1   Introduction

The Mann–Whitney test is a **two-sample** test – the non-parametric equivalent of the two-sample *t*-test (10.2). Due to the strong parallels between the Mann–Whitney test and the two-sample *t*-test, it would be useful for the readers first to review the use of *t*-tests.

12.2.2   Test statistics, *W* and *U*

The Mann–Whitney test compares the median values of two samples. The procedure for this test is illustrated using Example 12.4.

Example 12.4

Is there a significant difference between the median values of the populations from which the following data sets, A and B, were drawn?

Set A	21	23	24	18	20	25
Set B	18	16	20	18	22	

The analysis is performed in the following text.

We can plot the raw data using box and whisker plots, as in Figure 12.4.

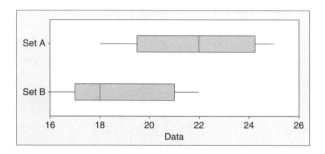

Figure 12.4   Boxplots of data in Example 12.4 (using Minitab).

The hypotheses for the 2-tailed Mann–Whitney test in Example 12.4 are:

Proposed hypothesis, H_1: $m_A \neq m_B$

Null hypothesis, H_0: $m_A = m_B$

where m_A and m_B are the true median values of populations A and B.

We choose the significance level $\alpha = 0.05$.

The first step is to rank the data values in order, keeping a record of the data set to which each value belongs – see Table 12.4. Where two or more data items have the same value, then they are given the same average rank, e.g. the three data items that have the value '18' share ranks '2', '3' and '4', giving the *average* rank '3' for all of them.

Table 12.4.   Ranking data for Example 12.4.

Set	B	A	B	B	A	B	A	B	A	A	A
Value	16	18	18	18	20	20	21	22	23	24	25
Rank	1	3	3	3	5.5	5.5	7	8	9	10	11

The sums of ranks are calculated for each set separately:

$$W_A = 3 + 5.5 + 7 + 9 + 10 + 11 = 45.5$$

$$W_B = 1 + 3 + 3 + 5.5 + 8 = 20.5$$

The test statistic is the **U-statistic** that is calculated for each set, x:

$$U_x = W_x - n_x(n_x + 1)/2 \qquad [12.2]$$

where n_x is the number of data values in set x.

In Example 12.4, we get

$$U_A = 45.5 - 6 \times (6 + 1)/2 = 24.5$$
$$U_B = 20.5 - 5 \times (5 + 1)/2 = 5.5$$

If the sample medians, m_A and m_B, are similar (null hypothesis), we would also expect the values of U_A and U_B to be similar. However, for the proposed hypothesis to be true, we would expect there to be a significant difference between U_A and U_B, with *one of them* having a significantly *low value*.

The critical values normally given for the Mann–Whitney test are the lower critical values, U_L. Appendix VI gives critical values at $\alpha = 0.5$ for sample sizes n_A and n_B.

The conditions for accepting the proposed hypothesis, H_1, are:

$$\text{Accept } H_1 \text{ if } \quad \textit{either } U_A \leqslant U_L \textit{ or } U_B \leqslant U_L.$$

In Example 12.4, the lower critical value, U_L, for a 2-tailed test with $n_A = 6, n_B = 5$ and $\alpha = 0.5$ is $U_L = 3$.

Since $U_A > U_L$ *and* $U_B > U_L$, we do not accept H_1.

A 2-tailed Mann–Whitney test for the problem in Example 12.4 is performed on the Website using Minitab and returns a p-value of 0.097. Since $p > 0.05$, we do not accept H_1, which agrees with the analysis above.

12.2.3 Applying the two-sample Mann–Whitney test

Example 12.5

Referring to the data in Example 12.1 presented in Chapter 12 Overview, we wish to use 2-tailed Mann–Whitney tests to investigate whether there are significant differences in the true medians between the responses to each pair of statements S1/S2, S1/S3 and S2/S3 respectively.

The analysis is performed in the following text.

The relevant 2-tailed hypotheses for the question in Example 12.5 would be of the form:

$$H_1: \quad m_X \neq m_Y$$

$$H_0: \quad m_X = m_Y$$

Using Minitab on the Website to perform a 2-tailed Mann–Whitney test we obtain the values in Table 12.5 for the lower U value, U_L, and p-value:

Table 12.5. Results for Example 12.5.

Comparison pairs	W	U	U_L	p-value	Decision	Conclusion
S1 and S2	50	14	13	0.0572	Do not accept H_1	m_{S1} could equal m_{S2}
S1 and S3	59	23	13	0.3571	Do not accept H_1	m_{S1} could equal m_{S3}
S2 and S3	41	5	13	0.0042	Accept H_1	$m_{S2} \neq m_{S3}$

It appears that there is a clear difference between the opinion scores for statements S2 and S3. The result for the S1/S2 pair fails to provide sufficient evidence that there is a difference between these two statements. However, in this calculation we have not considered the fact that some students may record generally higher or lower opinion scores than the other students – see Figure 12.1. The variation *between students* may mask a genuine difference *between statements* – refer to the use of the *paired* Wilcoxon test which will be considered in Example 12.7.

Q12.2

The generation times (5.2.4), t, of cultures of the same micro-organisms were recorded under two conditions X and Y and reproduced in the table below:

Condition X	6.7	5.8	6.9	9.6	8.9	8.2	6.1	4.8	9.2
Condition Y	3.9	6.3	4.4	5.6	6.3	4.2	7.2		

Is there evidence that the generation times are significantly different under the two conditions?

12.3 Paired Wilcoxon Test

12.3.1 Introduction

The paired Wilcoxon test is the non-parametric equivalent of the paired t-test (10.3). Due to the strong parallels between the paired Wilcoxon test and the paired t-test, it would be useful for the readers first to review the use of paired t-tests.

12.3.2 Test statistic, W

The one-sample Wilcoxon test forms the basis of this *paired* test, in a similar way to the one-sample *t*-test which forms a basis for a paired *t*-test (10.3).

A paired test is performed between two samples, in which each data value in one sample can be 'paired' uniquely with one data value in the other sample.

To perform the non-parametric paired test, it is first necessary to calculate the differences between each pair of values, and then the one-sample Wilcoxon test is used to compare those *differences* with a 'null' median value of zero.

The procedure for this test is illustrated using Example 12.6.

Example 12.6

The reaction times of nine subjects, before and after being given a particular drug, are recorded below.

The differences between the reaction times have also been calculated, together with the ranking of these differences.

Subject	1	2	3	4	5	6	7	8	9
Before	15.5	16.6	24.0	11.8	14.5	19.4	14.7	23.1	22.7
After	17.6	17.5	25.3	10.8	17.6	19.4	17.2	26.1	20.6
Differences	2.1	0.9	1.3	−1.0	3.1	0	2.5	3.0	−2.1
Ranks (ignoring sign)	4.5	1	3	2	8		6	7	4.5
Sign	+	+	+	−	+		+	+	−

Is there sufficient evidence to show that the drug has affected reaction time?

The analysis is performed in the following text.

The hypotheses for the 2-tailed paired Wilcoxon test are therefore:

Proposed hypothesis, H_1: $m_{AFTER} \neq m_{BEFORE}$

Null hypothesis, H_0: $m_{AFTER} = m_{BEFORE}$

The differences between each pair of the data values in Example 12.6 are ranked using the conditions that:

- Pairs with *zero difference* should be excluded from further calculations.
- Pairs with *equal differences* should share the ranking values.

The test statistics can be calculated for Example 12.6:

$$W(+) = 4.5 + 1 + 3 + 8 + 6 + 7 = 29.5$$
$$W(-) = 2 + 4.5 = 6.5$$

In Example 12.6, the hypothesis test is 2-tailed, and if we assume that $\alpha = 0.5$, the critical value for $n = 8$ is $W_L = 3$.

Since both $W(-) > W_L$ and $W(+) > W_L$ we do not accept the proposed hypothesis.

A 2-tailed paired Wilcoxon test for the problem in Example 12.6 is performed on the Website using Minitab and returns a p-value of 0.123. Since $p > 0.05$, we do not accept H_1, which agrees with the analysis above.

12.3.3   Applying the paired Wilcoxon test

Example 12.7

Referring to the data in Example 12.1 presented in Chapter 12 Overview, we wish to use a paired Wilcoxon test to test whether there is a difference between S1 and S2, taking the students as being the unique factor that defines each data pair.

The analysis is performed in the following text.

Table 12.6 contains the differences between the results S1 and S2 for each student.

Table 12.6.   Differences in the responses from individual students.

Student	a	b	c	d	e	f	g	h
S1−S2	2	1	2	2	2	0	0	1

Using Minitab on the Website to perform a one-sample Wilcoxon test we obtain the values in Table 12.7.

Table 12.7.   Results for Example 12.7.

Difference	$W(-)$	W_L	p-value	Decision	Conclusion
S1−S2	0	0	0.036	Accept H_1	$m_{S1} \neq m_{S2}$

It is important to compare the 'significant' result obtained by using a paired test for this problem with the 'not significant' result obtained in Example 12.5 when using a Mann–Whitney test to compare the results for S1 and S2. The paired test is more powerful (9.4.5) because it uses the extra information linking data differences to individual students.

Q12.3

This is the same as Q10.6, but assumes that the data may not necessarily be normally distributed.

Seven 'experts' in the taste of real ale have been asked to give a 'taste' score to each of two brands of beer, Old Whallop and Rough Deal. The results, a_i and b_i, for each expert are as follows, together with the differences, d_i, between the scores for each expert:

Experts	A	B	C	D	E	F	G
Old Whallop, a_i	60	59	65	53	86	78	56
Rough Deal, b_i	45	62	53	47	65	80	46
Difference, $d_i = a_i - b_i$	15	−3	12	6	21	−2	10

Perform a paired Wilcoxon test on the two data sets to test whether their medians are significantly different.

12.4   Kruskal–Wallis and Friedman Tests

12.4.1   Introduction

The Mann–Whitney test (12.2) looked for differences in median values between just *two* samples. Both the Kruskal–Wallis and Friedman tests look for differences in median values between *more than two* samples.

The Kruskal–Wallis test is used to analyse the effects of more than two levels of just *one factor* on the experimental result. It is the non-parametric equivalent of the one-way ANOVA (11.1).

The Friedman test analyses the effect of two factors, and is the non-parametric equivalent of the two-way ANOVA (11.2).

12.4.2   Test statistics, *H* and *S*

The calculations for the Kruskal–Wallis and Friedman test statistics, H_{STAT} and S_{STAT} respectively, are based on the ranking of individual data values. These calculations are given on the Website.

12.4.3 Applying the Kruskal–Wallis test

Example 12.8

Referring to the data in Example 12.1 presented in Chapter 12 Overview, we wish to use a single Kruskal–Wallis test to test whether there are any significant differences in the true medians between the responses to all the *statements* (S1, S2 and S3).

The analysis is performed in the following text.

In Example 12.5 we used three Mann–Whitney tests to check for any difference between the three statements, but here we aim to use just the one Kruskal–Wallis test.

Using Minitab on the Website to perform the Kruskal–Wallis test with 'statement' as the *factor* gives us the results in Table 12.8.

Table 12.8. Results of Kruskal–Wallis test for 'statement' as a factor.

Factor/levels	p-value	Decision	Conclusion
Statements S1, S2, S3	0.012	Accept H_1	Difference(s) exist between statement medians

The Kruskal–Wallis test is similar to the one-way ANOVA in that it will identify that a difference exists, but then we would have to use a *post hoc* investigation to decide where the differences between the samples might lie.

Example 12.9

Referring again to the data in Example 12.1 presented in Chapter 12 Overview, we will investigate whether the Kruskal–Wallis test will identify any differences between the median responses of *individual students* (a, b, c, d, e, f, g, h).

The analysis is performed in the following text.

Using Minitab on the Website to perform the Kruskal–Wallis test with 'student' as the *factor* gives us the results in Table 12.9.

Table 12.9. Results of Kruskal–Wallis test for 'student' as a factor.

Factor/levels	p-value	Decision	Conclusion
Students a, b, c, d, e, f, g, h	0.142	Not accept H_1	No evidence of differences

The test has been unable to detect any significant differences between the students, although it appears from Figure 12.1 that there is some considerable variation between their median values.

The problem here is that the variation of each student's response to the three statements has masked any trend between the students. We will see that it is necessary to use the two-way Friedman test (12.4.4) to resolve the problem. Compare this situation with Example 11.3 and the use of one-way and two-way ANOVAs.

Q12.4

The fungi species richness was measured on a 10-point scale on four different species of trees over a period of 4 days, and the results tabulated as below:

	Day 1	Day 2	Day 3	Day 4
Tree 1	6	4	3	3
Tree 2	4	3	3	2
Tree 3	4	2	1	1
Tree 4	2	1	2	1

Perform a Kruskal–Wallis test to investigate whether there is a significant difference in fungi species richness between the trees. Take the measurements on different days as being replicate measurements.

12.4.4 Applying the Friedman test

The Friedman test is similar to a two-way ANOVA in that it will take *account of* the effect of two possible factors at the same time. However, in each calculation, the Friedman test will only evaluate the significance of one, *main*, factor, by dividing the data into 'blocks' on the basis of the second, *blocking*, factor. The significance of the second factor can be found in a *separate* calculation by reversing the roles of the two factors.

Example 12.10

Referring to the data in Example 12.1 presented in Chapter 12 Overview, we wish to use a single Friedman test to test whether there are any significant differences in the true medians between the students while blocking the data according to the responses to the separate statements.

The analysis is performed in the following text.

Using Minitab on the Website to perform the Friedman test with the 'student' as the *main factor* and 'statement' as the *blocking factor*, we find:

Factor	Blocking	p-value	Decision	Conclusion
Student	Statement	0.020	Accept H_1	Difference(s) exist between student medians

By taking into account the variation between statements, this test has now been able to detect a significant difference between the students. This was not possible (see Example 12.9) by using the Kruskal–Wallis test alone.

Q12.5

Use the same data as in Q12.4 for fungi species richness:

	Day 1	Day 2	Day 3	Day 4
Tree 1	6	4	3	3
Tree 2	4	3	3	2
Tree 3	4	2	1	1
Tree 4	2	1	2	1

Perform a Friedman test to investigate whether there is a significant difference in fungi species richness between the trees (treatment), while blocking the data by 'day'.

Q12.6

Use the same data as in Q12.4 and Q12.5 for fungi species richness.

Perform a Friedman test to investigate whether there is a significant difference in fungi species richness between the days, while blocking the data by 'tree'.

13

Correlation and Regression

Overview

- 'How to do it' video answers for all 'Q' questions.
- Correlation tests in Excel and Minitab.
- Excel tutorial: uncertainty in linear calibration.

Correlation is a measure of the extent to which the values of two variables of a system are *related* (or correlated). For example, both the height and weight of a child normally increase as the child gets older – there is a correlation between height and weight.

It must be noted, however, that the fact that two parameters are correlated does *not necessarily* mean that one is a *cause* and the other an *effect*. In the example of the height and weight of a child, both increases are the result of the growth of the child. The underlying 'cause' (factor) is the maturity of the child, and the height and weight are *both* 'effects' (outcomes).

If two variables are correlated, the relationship between them could follow very many different mathematical functions, e.g. the numbers of bacteria in a dying population may follow an exponential decay equation. However, *linear correlation* is a measure of the extent to which there is a linear relationship between two variables, i.e. the extent to which their relationship can be expressed in terms of a *straight line*.

A *linear regression* analysis is the process of deriving the slope and intercept of the *straight line* equation that can be used to describe the relationship between the two variables. When performing a linear regression it is normally assumed that there is a causal relationship, i.e. the value of one variable is dependent on the value of the other. We saw in 4.1.2 that the dependent variable is placed on the y-axis of an $x-y$ graph with the independent variable on the x-axis.

In simple terms:

- 'correlation' asks *if* there is a relationship between two variables; and
- 'regression' measures *how* one variable is dependent on the other through the slope and intercept of a 'best-fit' straight line.

Essential Mathematics and Statistics for Science 2nd Edition   Graham Currell and Antony Dowman
Copyright © 2009 John Wiley & Sons, Ltd

In this chapter, we first introduce the correlation coefficient (13.1) as the test statistic for correlation before developing the statistics of regression and correlation (13.2). We then (13.3) use the statistics to develop calculations for the experimental uncertainty that arises from using a linear calibration line.

13.1 Linear Correlation

13.1.1 Introduction

The correlation statistic measures how accurately a change in one variable may be predicted by the change in another variable.

For example, the mass of a pure copper bar can be predicted very accurately by its volume, i.e. near-perfect correlation between mass and volume, depending only on the accuracy of the measurement conditions. However the mass of an adult man cannot be predicted so accurately from his volume, because of variations in body structures and densities, i.e. some correlation between mass and volume, but certainly not perfect.

13.1.2 Linear correlation coefficient

The **linear correlation coefficient**, r, between two variables, x and y, is a measure of the extent to which the data follows a straight line relationship of the form $y = mx + c$, where m is the slope and c is the intercept [4.5].

The *statistic*, r, is also called the 'product-moment correlation coefficient' or **Pearson's** correlation coefficient. The use of r as a statistic in a hypothesis test assumes a normal distribution in the uncertainty of the y data values (8.1.3).

For ordinal data, or data from non-normal distributions, **Spearman's rank** correlation coefficient, r_S, is based on the ranking of the data values – see other *non-parametric* tests in Chapter 12.

The square of Pearson's correlation coefficient, r^2, is called the **coefficient of determination** and is also often used as a measure of correlation.

Each of the graphs in Figure 13.1 shows the best-fit straight line drawn through five data points:

- If the data values for x and y fall *exactly* on straight lines as in (a), (b) and (c), there is *perfect* linear correlation, $r^2 = 1$, and either $r = +1$ or $r = -1$.
- Note that the *sign* of r is the same as the *sign* of the slope of the line, e.g. r is positive for graphs (a), (b) and (e), but negative for (c) and (f).
- However, the magnitude of r does *not* depend on the *slope* of the line, e.g. graphs (a) and (b) have the same value of r but different slopes.
- Where the data points do not fall exactly on the straight line, then the magnitude of r lies between -1 and $+1$, e.g. graphs (e) and (f).
- If the 'best-fit' straight line has zero slope, then $r = 0$ as in (d).

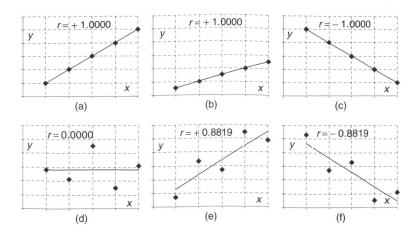

Figure 13.1   Data relationships.

The relationship can be summarized as in Table 13.1.

Excel calculates the correlation coefficient using functions CORREL and PEARSON (Appendix I).

Table 13.1.   Ranges of the correlation coefficient.

Relationship	Slope	Correlation coefficient	Coefficient of determination
	m	r	r^2
Perfect linear correlation	Positive	$+1$	$+1$
Correlation plus random variation	Positive	$0 < r < +1$	$0 < r^2 < +1$
No correlation	Zero	0	0
Correlation plus random variation	Negative	$-1 < r < 0$	$0 < r^2 < +1$
Perfect linear correlation	Negative	-1	$+1$

Example 13.1

The best-fit straight line for the data in Figure 13.2 is a horizontal line, with zero slope and a correlation coefficient equal to zero: $r = 0$.

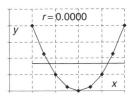

Figure 13.2

Using the above information, can you say that the data in Figure 13.2 is correlated?

The fact that $r = 0$ is a clear statement that there is no *linear* correlation in the data. However, simple observation of the data does show a very definite *nonlinear* correlated relationship between x and y. In fact y is proportional to x^2.

From Example 13.1, we can see that it is important to remember that the correlation coefficient *only* records the significance of a *linear* term in a relationship. It is *always* necessary to plot the data (e.g. in Excel) and *examine the data by eye* to see whether there may be a *nonlinear relationship*, before deriving the correlation coefficient (13.3.2).

13.1.3 Hypothesis test for linear correlation

The hypothesis test for linear correlation between variables x and y will test whether a best-fit straight line (4.2) will give a slope that is not zero (for a 2-tailed test), or sloping in a particular direction (for a 1-tailed test).

The basic hypotheses for a 2-tailed test would be:

Proposed hypothesis, H_1: Slope of y against x is not zero, $m \neq 0$.

Null hypothesis, H_0: Slope of y against x is zero, $m = 0$.

The test statistic for correlation is the value of the correlation coefficient, r.

Statistics software will usually also provide a p-value for a correlation test, in addition to calculating the correlation coefficient, r.

The critical value, r_{CRIT}, can be derived from the table of Pearson's correlation coefficients (Appendix V), using appropriate values for:

T, number of tails;

α, the significance level required (typically $\alpha = 0.05$); and

df, the degrees of freedom, $df = n - 2$ (where $n =$ the number of data pairs).

The proposed hypothesis would be accepted if the positive value of r were greater than or equal to the critical value:

$$\text{Accept } H_1 \text{ if } |r| \geq +r_{CRIT} \qquad [13.1]$$

Example 13.2

The data below shows the number of 'press-ups' and 'sit-ups' achieved by seven children (subjects). From experience of previous experiments, we devise the hypothesis that a more athletic child who can perform more 'press-ups' will also be able to perform more 'sit-ups'. Hence we investigate whether the data shows a *1-tailed* correlation between the performances in the two activities.

Subject	1	2	3	4	5	6	7
Press-ups	10	8	2	6	7	3	5
Sit-ups	24	27	12	14	21	16	22

The analysis is performed in the following text.

For Example 13.2, a 1-tailed test for correlation at a significance level of 0.05 gives the values in Table 13.2.

Table 13.2. Results for Example 13.2.

r	r_{CRIT}	p-value	Decision	Conclusion
0.781	0.669	0.038	Accept H_1	The numbers of press-ups and sit-ups show positive linear correlation

Q13.1

An investigation was conducted into whether students' performances in music are related to their performances in mathematics. The data given in the table shows the performance scores of a group of eight students in both mathematics and music.

Student	1	2	3	4	5	6	7	8
Maths mark	65	53	71	63	49	58	54	73
Music mark	59	47	61	65	52	65	51	64

(i) Use Excel to calculate the correlation coefficient for the above data.
(ii) Compare the result in (i) with the critical value for a 1-tailed test for positive correlation between music and mathematics scores.
(iii) Use statistics software to calculate the appropriate p-value, and compare with the conclusion reached in (ii).

13.2 Statistics of Correlation and Regression

13.2.1 Introduction

It is useful to consider again the difference between testing for correlation and performing a regression calculation:

- The **correlation calculation** tests whether two variables might show a linear relationship with a line of best fit which has a non-zero slope.
- The **regression calculation** is performed after it has been accepted that there is correlation between the variables and the regression then calculates the actual values of the best-fit slope and intercept.

The use of a best-fit straight line was introduced in 4.2. In this unit we develop the statistics that calculate the slope and intercept of this line, and then proceed to see how this leads to a correlation test for a non-zero slope.

13.2.2 Statistics of linear regression of 'y on x'

The process of linear regression calculates the coefficients of the best-fit straight line to a set of 'linear' data (4.2).

The best-fit line of regression of y on x can be described by:

$$y = mx + c$$

where the regression coefficients are:

$m = slope$ (also called the *coefficient* of x)

$c = intercept$

For the 'regression of y on x', it is assumed that the only uncertainties in the data are in the measurement of the y-values. Hence it is important that the measured variable with the greatest uncertainty is placed on the y-axis; this is normally the *dependent* variable. Reversing the data would give slightly *different* results.

The **residual, R**, for a given data point, (x, y), is the difference in the y-values between the data point (y) and the point on the best-fit straight line (y') at the same x-value - see Figure 13.3.

In general, for a point (x_i, y_i), the residual is given by:

$$R_i = y_i - y_i' \qquad [13.2]$$

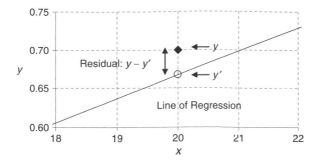

Figure 13.3 Residual on a best-fit line (using data from Example 13.4).

Example 13.3

A data point with co-ordinates (20, 0.7) in Figure 13.3 is part of a data set (see Example 13.4) that has a best-fit line of linear regression given by the equation:

$$y = 0.0308x + 0.051$$

Calculate:

(i) The y-value, y', of the point on the line of regression at the same value of x.
(ii) The 'vertical' difference between the data point and the line of regression.

(i) Substituting $x = 20$ into the equation of the line gives:

$$y' = 0.0308 \times 20 + 0.051 = 0.667$$

(ii) The 'vertical' difference is:

$$Residual, R = y - y' = 0.700 - 0.667 = +0.033$$

The **residual sum of the squares** (SS_{RESID}) is defined by taking the sum of the squares of the residuals for all the data points:

$$SS_{RESID} = \sum_i R_i^2 \qquad [13.3]$$

A regression calculation in software works by adjusting the slope, m, and intercept, c, of the straight line to obtain a *minimum* value for the 'residual sum of the squares'. The best-fit straight line has the *smallest possible* value for SS_{RESID}.

For the above reason, the process of finding the best-fit straight line is also described as the **method of least squares**.

Once the best fit has been achieved, residual differences in y-values still exist between each data point and the line of regression. The overall residual difference in the 'fit' is quantified by the **standard error of regression**, SE_{yx}:

$$SE_{yx} = \sqrt{\frac{SS_{RESID}}{n-2}} \qquad [13.4]$$

The standard error of regression is a 'best estimate' for the *standard deviation of the experimental uncertainty*, σ_E (11.1.2).

The standard error of regression can be calculated directly in Excel by using the STEYX function.

Example 13.4

In an experiment to find the concentration, C_o, of an unknown chemical solution, the absorbances, A, of four standard solutions of known concentrations were measured in a spectrophotometer, recording the values in the table below. Plotting the data on an $x-y$ graph, the y variable is the absorbance, A, and the x variable is the concentration, C.

The slope and intercept of the best-fit straight line are calculated using the SLOPE and INTERCEPT functions in Excel, giving $m = 0.0308$ and $c = 0.051$ respectively.

For each value of C, calculate the values of:

(i) A' (the y-value for the point on the calibration line); and
(ii) $A - A'$ (the residual for that point).
    Calculate the values of:
(iii) SS_{RESID}
(iv) SS_{YX}

C	(x)	10	15	20	25
A	(y)	0.37	0.48	0.7	0.81

(i) Each value of A' is calculated using the equation $A' = m \times C + c$:

	A'	(y')	0.359	0.513	0.667	0.821
(ii)	$A - A'$	$(y - y')$	0.011	-0.033	0.033	-0.011
	$(A - A')^2$	$(y - y')^2$	0.121×10^{-3}	1.089×10^{-3}	1.089×10^{-3}	0.121×10^{-3}

(iii) $SS_{RESID} = \sum (A - A')^2 = 0.00242$

(iv) $SE_{yx} = \sqrt{\dfrac{SS_{RESID}}{n-2}} = \sqrt{\dfrac{0.00242}{4-2}} = 0.0348$

13.2.3 Statistics of linear correlation

In this section we give a simplified approach to the statistics of correlation. A more in-depth approach is given on the Website.

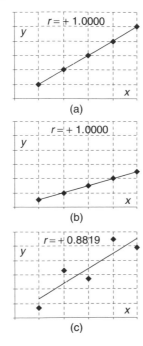

Figure 13.4   Comparative correlation.

In Figure 13.4, the x-data has the same values in each of the three graphs, hence the sample standard deviation, s_x, is also the same in each case.

In graphs (a) and (b) the data points are *perfectly correlated* on the straight line, so that the standard deviation, s_y, of the data in the y-direction will be directly related to the standard deviation, s_x, of the data in the x-direction, and dependent on the slope of the line, i.e. for perfect correlation s_y can be *predicted* using the equation:

$$s_y = m \times s_x \qquad [13.5]$$

Equation [13.5] shows that the 'spread' of the y-values, s_y, will be greater for lines with a larger slope – as can be seen in moving from graph (a) to (b).

If we now add additional uncertainty in the y-values, as in (c), the spread of y-values will increase, such that s_y will be *greater* than $m \times s_x$.

We can now define the **correlation coefficient**, r, as the *proportion of the variation in y predicted by the variation in x*, giving equation [13.6]:

$$r = \frac{m \times s_x}{s_y} \qquad [13.6]$$

For perfect correlation in (a) and (b), $s_y = m \times s_x$, giving $r = 1.0$.

With additional uncertainty, as in (c), $s_y > m \times s_x$, making $|r| < 1.0$.

Q13.2

The following data shows the distances for a standing jump achieved by 10 students, whose heights are recorded:

Student	1	2	3	4	5	6	7	8	9	10
Height (m)	1.59	1.63	1.67	1.7	1.72	1.8	1.74	1.75	1.82	1.84
Jump (m)	1.49	1.45	1.78	1.6	1.9	1.55	1.69	1.81	1.69	1.93

Performing a linear regression analysis on the data gives:

Slope of best-fit line, $m = 1.11$

Sample variance of height data, $s_x^2 = 0.006627$

Sample variance of jump data, $s_y^2 = 0.027721$

(i) Calculate Pearson's correlation coefficient between the two data sets.
(ii) Compare the result in (i) with the critical value for a 1-tailed test for positive correlation between jump distance and height.
(iii) Use statistics software to calculate the appropriate p-value, and compare with the conclusion reached in (ii).

13.3 Uncertainty in Linear Calibration

13.3.1 Introduction

The use of linear best-fit calibration lines is a very common procedure in experimental measurements. However, it is also common that the possible errors or uncertainty in the result of a given calculation are often omitted.

In this unit we first use a plot of residuals to confirm that there is no underlying curvature in the calibration data, and then introduce the equation that can be used to estimate the uncertainty in a calculated value.

A more in-depth examination of the statistics involved in these topics is given on the Website.

13.3.2 Curvature in calibration – checking residuals

When the data is being used for a linear calibration graph (4.2.6), it is normal to expect a high level of linear correlation, with (typically) $r^2 > 0.99$. However, a high value of r^2 can still hide significant nonlinear characteristics that might indicate an inherent problem with the 'linear' calibration process.

The **residual** for each data point is the deviation (in the y-direction) between the point itself and the best-fit calibration line (13.2.2). In an experiment involving linear calibration, it is advisable to check the residuals arising from the best-fit straight line. Example 13.5 illustrates how a simple plot of residual values highlights inherent curvature within a 'linear' calibration.

Example 13.5

Two 'linear' calibration graphs, A and B, are shown in Figure 13.5(a) and Figure 13.5(b) respectively. Both show good linear correlation with $r^2 = 0.991$.

The 'residuals' for every data point in the two lines are shown (magnified) in the graphs underneath: Figure 13.6(a) and Figure 13.6(b) respectively.

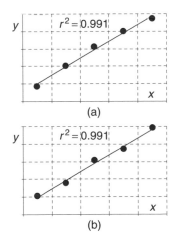

Figure 13.5 (a) Calibration A. (b) Calibration B.

What information do the 'residual' plots provide?

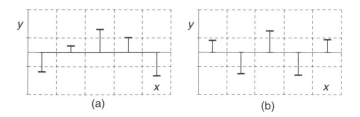

Figure 13.6   (a) Residuals A. (b) Residuals B.

The residuals for line A show a 'systematic' variation along the line, clearly indicating that the 'true' line has a distinct curvature. The 'less than perfect correlation' for line A is due mainly to the curvature of the line and not to measurement uncertainty. The experimenter should investigate the reason for this apparent curvature before proceeding with the experiment.

The residuals for line B show 'random' variations consistent with experimental uncertainty. The 'less than perfect correlation' for line B is due mainly to experimental uncertainty which can then be taken into account (13.2.2) when calculating the overall uncertainty in the final result of the experiment.

Q13.3

Analyse the linear correlation of the following data sets and comment on their use as possible calibration graphs (the dependent variable is in the second row of each data set):

(i)

x	1	3	5	7	9	11
y	0.60	1.77	2.80	3.85	4.77	5.55

(ii)

T	0.1	0.3	0.5	0.7	0.9	1.1
V	0.07	0.18	0.31	0.46	0.57	0.73

(iii)

C	1	3	5	7	9	11
A	0.06	0.18	0.30	0.42	0.52	0.59

13.3.3 Uncertainty in linear calibration

Example 13.6

In an experiment to find the concentration, C_0, of an unknown chemical solution, the absorbances, A, of four standard solutions of known concentrations were measured in a spectrophotometer, recording the values (as in Example 13.4) below.

Concentration, C	(x)	10	15	20	25
Absorbance, A	(y)	0.37	0.48	0.70	0.81

Three replicate measurements were made of the absorbance of the unknown solution, obtaining an average value of $A_0 = 0.57$:

  (i) Use a best-fit calibration line for the standard solutions' data to obtain a best estimate of the value of C_0 corresponding to the measured value of A_0.
 (ii) Estimate the uncertainty in the value of C_0 obtained in (i).
(iii) Calculate the 95% confidence interval for the true value of C_0.

Answers are given in the following text.

Example 13.6 illustrates a typical calibration procedure, in which we aim to calculate the **unknown value of a 'test sample'** (e.g. its chemical concentration, C_0) by comparison with the known values of a number of prepared 'standard' samples.

We assume that the sample being measured has a property '*y*' **that varies linearly with the value '*x*'** that we are trying to find. For example, the absorbance, '*A*', of a chemical solution often varies linearly with its concentration, '*C*' (Beer–Lambert law).

We start with a number of **standard samples with known x-values** (e.g. known concentrations, C), measure the *y*-values (e.g. absorbances, A) of the standard samples, and plot the data points on the x–y (or A–C) graph, together with a best-fit calibration line – see Figure 13.7.

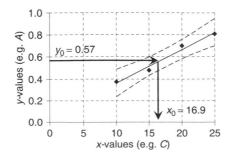

Figure 13.7 Using a calibration line for Example 13.6.

We then measure the y–value, $\boldsymbol{y_0}$ (e.g. $\boldsymbol{A_0}$) **of the test sample**, and use the calibration line to find the equivalent x-value, $x_0(C_0 = 16.9)$, by reading the value of x_0 ($= 16.9$) equivalent to y_0 ($A_0 = 0.57$) – see the arrows in Figure 13.7.

It is possible to use a *graphical* construction to obtain an estimate for x_0 (or C_0), as above. However, we now find the value of x_0 by calculation:

- Typically the best-fit calibration line is drawn with $\boldsymbol{n}$ *calibration data points*.
- The slope, $\boldsymbol{m}$, and intercept, $\boldsymbol{c}$, of the calibration line are calculated.
- A 'best-estimate' y-value, $\boldsymbol{y_0}$, of the unknown sample is calculated by taking the *mean value* of $\boldsymbol{k}$ *replicate sample measurements*.
- The 'best-estimate' x-value, $\boldsymbol{x_0}$, of the unknown sample is then calculated by using the equation:

$$x_0 = \frac{y_0 - c}{m} \qquad [13.7]$$

For Example 13.6:

$$x_0 = \frac{0.57 - 0.051}{0.0308} = 16.85$$

Having calculated the best estimate of the unknown, x_0, value, the next step is to calculate an estimate of the experimental uncertainty, u_{x_0}, in that value.

The uncertainty, u_{x_0}, arises from two sources:

- Uncertainty in the true position of the calibration line, based on n data points.
- Uncertainty in the true value of y_0, based on k replicate measurements.

In Figure 13.7 the dashed lines in the diagram show the ranges within which it is possible to be 95% confident of drawing the position of a best-fit calibration line as a 'free fit' of both slope *and* intercept.

For the 'free fit' calibration line in Figure 13.7, the uncertainty u_{x_0} is given by:

$$u_{x_0} = \frac{SE_{yx}}{m} \times \sqrt{\frac{1}{k} + \frac{1}{n} + \frac{(y_0 - \bar{y})^2}{m^2 \times (n - 1) \times s_x^2}} \qquad [13.8]$$

where:

- SE_{yx} is the standard uncertainty of regression – see 13.2.2;
- m is the slope of the calibration line, e.g. a 'flatter' slope (small m) would *increase* the range of uncertainty in the x-value for a given uncertainty in the y-value;
- $\bar{y}$ is the mean value of the y-data calibration values;

- s_x^2 is the sample variance (= standard deviation^2) of the x-data calibration values;
- n is the number of calibration data values;
- k is the number of replicate measurements made to calculate y_0.

The last term in the square root of [13.8] is the factor that is responsible for the 'opening up' of the uncertainty range at both ends of the calibration line in Figure 13.7. This term becomes *zero* (and disappears) when y_0 equals the mean y-value ($\bar{y}$) of the points used to generate the best-fit line, i.e. when the measured value falls in the 'middle' of the calibration values.

For *well-designed* experiments where the measured value of y_0 falls within the *middle half of the calibration line*, we can make the approximation that the last term in the square root will then be small, and can usually be omitted. This gives the approximate equation for the free fit calibration line:

$$u_{x_0} \approx \frac{SE_{yx}}{m} \times \sqrt{\frac{1}{k} + \frac{1}{n}} \qquad [13.9]$$

For Example 13.6 these values are:

$SE_{yx} = 0.0348, m = 0.0308$ (from Example 13.4)

$\bar{y} = 0.59,\ s_x^2 = 41.67$ (by calculation, e.g. in Excel)

$y_o = 0.57$

$n = 4, k = 3$

Then, entering these values into [13.8] we find:

$$u_{x_0} = \frac{0.0348}{0.0308} \times \sqrt{0.333 + 0.25 + 0.0034} = 0.865$$

We can see that, in this case, the last term in the square root is having negligible effect, being much smaller than the other terms in the square root. This is because our value for $y_0 = 0.57$ is very close to the middle of the calibration line, $\bar{y} = 0.59$.

If we just use equation [13.9] we get $u_{x_0} = 0.863$.

The standard uncertainties, u_{x_0}, calculated in [13.8] and [13.9] give a 68% confidence range for the unknown true value, μ.

The confidence deviation, $Cd_{\mu,X\%}$, for $X\%$ confidence (8.2.4) is given by:

$$Cd_{\mu,X\%} = t_{2,\alpha,df} \times u_{x_0} \qquad [13.10]$$

where the t-value is calculated for degrees of freedom $df = n - 2$ and $\alpha = 1 - X\%/100$, giving the confidence interval, $CI_{\mu, X\%}$:

$$CI_{\mu, X\%} = x_o \pm Cd_{\mu, X\%} \qquad\qquad [13.11]$$

For Example 13.6 these values are:

$$t-\text{value} \quad t_{2, \alpha, n-2} = 4.30$$

Using [13.10] $Cd = 4.30 \times 0.865 = 3.72$

Using [13.11] $CI = 16.85 \pm 3.72$

Hence there is a 95% probability that the true value lies between 13.13 and 20.57.

The uncertainty ranges due to calibration are often much wider than expected.

Example 13.6 showed a wide uncertainty range due to the relatively 'low' correlation in the calibration data, $r = 0.9899$. For a good calibration, many experimenters would expect a correlation with at least $r = 0.999$.

Q13.4

The data in the table below give the calibration for a spectrophotometric measurement, where the y variable is the absorbance, A, and the x variable is the concentration, C, of the standard solutions.

Concentration, C	(x)	0.5	1	1.5	2	2.5
Absorbance, A	(y)	0.13	0.27	0.30	0.49	0.53

Three replicates of the test solution give an average value $A_o = 0.36$.

Calculate:

(i) The best estimate of the concentration, C_o, of the test solution.
(ii) The standard uncertainty, u_{C_o}, of the concentration, C_o.
(iii) The 95% confidence interval of the concentration, C_o.

14

Frequency and Proportion

Overview

- 'How to do it' video answers for all 'Q' questions.
- Chi-squared tests in Excel and Minitab.
- Proportion tests in Minitab.

There are many occasions when the experimental variable being measured is the *frequency* with which particular events occur, e.g. counting how many seeds have germinated under different sets of conditions.

The chi-squared, χ^2, hypothesis tests provide a family of tests often used to analyse sets of frequencies. The calculated χ^2 statistic is used to test whether an observed distribution of frequencies might have occurred purely by chance or whether an underlying factor had caused a deviation from an expected distribution.

The first unit applies the χ^2 test to a contingency table which assesses whether differences between frequency distributions of observed events might be due to different external conditions. For example, we might investigate whether the number of seeds germinating is dependent on (contingent on) particular conditions of temperature and moisture.

The second unit tests whether a single set of frequencies represent a 'good fit' with an expected set of frequencies. The 'goodness of fit' test is used to test whether the observed distribution of events might be consistent with an underlying theory. This test can be used to assess whether observed distributions may, or may not, follow certain predicted distributions (e.g. Poisson).

When the possible observed events fall into one of just *two possible categories*, then we have a simple proportion, e.g. the proportion of times that a tossed coin will give a 'head'. We saw in 9.3.2 that the simple choice of two alternatives can be analysed using the binomial distribution, and in 14.3.2 this is developed into Fisher's exact test. We also investigate the use of both χ^2 statistics and a normal distribution when comparing an observed proportion with an expected proportion.

Excel functions can be used to perform a χ^2 test – see Appendix I. The Excel function CHIINV can also be used to give values for χ^2_{CRIT} (14.1.2 and 14.2.2).

Essential Mathematics and Statistics for Science 2nd Edition   Graham Currell and Antony Dowman
Copyright © 2009 John Wiley & Sons, Ltd

14.1   Chi-squared Contingency Table

14.1.1   Introduction

The chi-squared, χ^2, test is a hypothesis test which uses the *frequency of events* as the response variable being measured.

Perhaps the most common χ^2- test is a *contingency test* (or *contingency table*) which investigates whether the observed distribution of occurrences between categories depends on some external factor. This dependency can also be described as an *association* between the observed distribution and the given factor.

14.1.2   Chi-squared, χ^2, contingency table

The operation of the chi-squared, χ^2, contingency test is illustrated in Example 14.1, which aims to investigate whether the distribution of river creatures is *contingent on* (i.e. depends on) the particular river environment.

Example 14.1

An environmental scientist has investigated two rivers, the Exe and the Wye, and wants to find out whether the distribution of certain water creatures is the same in both. He recorded numbers of mayflies, sludge worms and hog lice in each river as in Table 14.1 below. He counted a grand total, T, of 150 creatures, with 'column' totals of 75 mayflies, 50 sludge worms and 25 hog lice. There were 'row' totals of 60 creatures in the Exe and 90 in the Wye, distributed as shown in the table:

Table 14.1.   Observed frequencies.

	Mayflies	Sludge worms	Hog lice	Row totals
Exe	24	21	15	$R_1 = 60$
Wye	51	29	10	$R_2 = 90$
Column totals	$C_1 = 75$	$C_2 = 50$	$C_3 = 25$	**Grand total: $T = 150$**

A simple inspection of the numbers might suggest that there are proportionately more hog lice in the Exe than the Wye and proportionately fewer mayflies. However, is there evidence, at a confidence level of 95 %, that this apparent difference in the distributions of the creatures within the two rivers did not occur just by chance?

The analysis is performed in the following text.

The first step is to calculate the numbers that would be *expected* in each category, *assuming* that the *total numbers* of creatures in the two rivers followed *exactly equal* distributions.

We start by noting values for:

- the row totals, $R_1(= 60)$ and R_2 $(= 90)$;
- the column totals, C_1 $(= 75)$, C_2 $(= 50)$ and $C_3(= 25)$; and
- the grand total, $T(= 150)$.

Notice that the row and column totals must *each* add up to 150 – this is a useful check for your additions.

We can see from the column totals that *overall* in the two rivers, the mayflies, sludge worms and hog lice are in the ratio of 75:50:25, which, for these convenient numbers, is the same as the simple ratio of 3:2:1.

Distributing the total of 60 creatures in the Exe according to this ratio, we would get 30 mayflies, 20 sludge worms and 10 hog lice. Similarly, distributing the total of 90 creatures in the Wye according to the *same distribution*, we would get 45 mayflies, 30 sludge worms and 15 hog lice. This calculation would give *expected frequencies* as in Table 14.2.

Table 14.2. Expected frequencies.

	Mayflies	Sludge worms	Hog lice	Row totals
Exe	30	20	10	$R_1 = 60$
Wye	45	30	15	$R_2 = 90$
Column totals	$C_1 = 75$	$C_2 = 50$	$C_3 = 25$	$T = 150$

Notice that the row and column *totals* are the same for both the 'observed frequencies' and the 'expected frequencies' – the only difference is that, in Table 14.2, the *row distributions* are the same and all the *column distributions* are the same.

The name 'expected' can be confusing. In reality we would *not* expect to see these 'expected' frequencies, because we would also expect *some* differences due to the randomness with which such events actually occur. The purpose of the chi-squared analysis is to test whether the differences between the observed frequencies and the 'ideal' expected frequencies are *too* great to be accounted for purely by chance randomness.

The relevant hypotheses for this chi-squared, χ^2, contingency test would now be:

Proposed hypothesis, H_1: The differences between observed and expected frequencies have not occurred purely by chance, and hence an additional factor is significant.

Null hypothesis, H_0: The differences between observed and expected frequencies could have occurred purely by chance.

We now derive a 'χ^2 statistic' to compare the *observed frequencies* with the *expected frequencies* that would occur if the distribution did *not* depend on the river. If the χ^2-test reports a significant difference, then we will conclude that the distribution *is* influenced by the choice of river.

The basic χ^2 statistic is calculated using the equation:

$$\chi^2 = \sum_i \frac{(O_i - E_i)^2}{E_i} \qquad [14.1]$$

where:

O_i is the observed frequency in category i, and
E_i is the expected frequency in category i.

We calculate the value of the χ^2 statistic for Example 14.1:

$$\chi^2 = \frac{(24-30)^2}{30} + \frac{(21-20)^2}{20} + \frac{(15-10)^2}{10} + \frac{(51-45)^2}{45} + \frac{(29-30)^2}{30} + \frac{(10-15)^2}{15}$$
$$= 6.25$$

In general, a *large* value for the χ^2 statistic would indicate that observed frequencies were *very different* from the expected frequencies. However, we need to know *how large* the value of the χ^2 statistic must be before we can accept that the difference is due to more than just random chance (i.e. accept the proposed hypothesis, H_1). Tables of critical values, χ^2_{CRIT} are given in Appendix III.

The number of degrees of freedom for a contingency table with r rows and c columns is given by:

$$df = (r-1) \times (c-1)$$

(We will see in 14.1.5, that if $df = 1$, we need to modify equation (14.1) and use the Yates correction.) For a 3×2 table, $df = (3-1) \times (2-1) = 2$. For a confidence of 95 %, the significance level $\alpha = 0.05$. Referring to Appendix III, for $df = 2$ and $\alpha = 0.05$, we find that $\chi^2_{CRIT} = 5.99$.

The decision on the hypothesis test can now be made:

$$\chi^2 \geqslant \chi^2_{CRIT} : \quad \text{Accept the proposed hypothesis}$$
$$\chi^2 < \chi^2_{CRIT} : \quad \text{Do not accept the proposed hypothesis}$$

In this example, the value of χ^2 is 6.25, giving:

$$\chi^2 > \chi^2_{CRIT}$$

Hence we conclude that there is *evidence that the pattern of frequencies differs from that which could be expected by chance* – we accept the proposed hypothesis.

The calculation for Example 14.1 is performed on the Website using both Excel and Minitab and returns a *p*-value equal to 0.044. On this basis ($p < 0.05$) we would again accept the proposed hypothesis that these results are not due to chance and that there is a significant difference between the distributions of creatures within the two rivers.

Q14.1

Look up the critical values, χ^2_{CRIT}, for the following conditions:

(i) 4 degrees of freedom at a significance level of 0.05;
(ii) 1 degree of freedom at a significance level of 0.01.

Q14.2

A χ^2 test records a value of $\chi^2 = 3.6$. Is it possible to say, with a confidence level of 95 % and degrees of freedom $df = 3$, that the differences between observed and expected frequencies did not occur just by chance?

14.1.3 Calculation of 'expected frequencies'

The numbers in Example 14.1 occurred in easy ratios, providing very easy calculations. However, we can also use a simple formula to calculate the frequencies in each cell in Table 14.2. The expected frequency, E_{xy}, for the cell in column x and row y is given by:

$$E_{xy} = \frac{C_x \times R_y}{T} \tag{14.2}$$

where C_x is the column total for column x and R_y is the row total for row y.

For example, for the number of sludge worms within the Exe, $x = 2$ and $y = 1$:

$$E_{21} = \frac{C_2 \times R_1}{T} = \frac{50 \times 60}{150} = 20$$

Example 14.2

In a double blind drug trial, 200 randomly selected patients who are all suffering from a specific disease are given either the drug or a placebo, and the numbers of each group who show no improvement, some improvement or much improvement are recorded and reproduced in Table 14.3. The row and column totals have also been calculated:

Table 14.3. Observed frequencies.

	Drug	Placebo	Row totals
Much improvement	26	13	$R_1 = 39$
Some improvement	56	36	$R_2 = 92$
No improvement	32	37	$R_3 = 69$
Column totals	$C_1 = 114$	$C_2 = 86$	$T = 200$

Calculate the expected frequencies assuming that there was no effect due to the drug.

The analysis is performed in the following text.

To calculate the number of people expected to show 'some improvement' with the 'drug' we use equation (14.2) to combine the first column total, $C_1 = 114$, the second row total, $R_2 = 92$, and the grand total, $T = 200$:

$$E_{12} = \frac{C_1 \times R_2}{T} \Rightarrow \frac{114 \times 92}{200} \Rightarrow 52.44$$

Note that it is acceptable to have non-integer values for the ideal expected frequencies.

Performing similar calculations for the other column/row combinations we obtain the results given in Table 14.4.

Table 14.4. Expected frequencies for Example 14.2.

	Drug	Placebo	Totals
Much improvement	22.23	16.77	$R_1 = 39$
Some improvement	52.44	39.56	$R_2 = 92$
No improvement	39.33	29.67	$R_3 = 69$
Totals	$C_1 = 114$	$C_2 = 86$	$T = 200$

Q14.3

Complete the contingency test calculation for the observed and expected frequencies calculated in Example 14.2.

Q14.4

Two students, A and B, are performing similar projects to investigate whether there is an association between the leaf colour (green, yellow/green, yellow) of seedlings and the soil in which they grow.

Of those seeds that germinate in each of three soil types, the students count the number of seedlings that fall into each of the categories for leaf colour. The results are given in the tables below.

Do both students arrive at the same conclusion concerning the suggestion that there is an association between leaf colour and soil type?

	Student set A			Student set B		
	Soil 1	Soil 2	Soil 3	Soil 1	Soil 2	Soil 3
Green	63	79	60	56	79	86
Yellow/green	20	25	19	11	25	19
Yellow	4	11	19	6	10	8

14.1.4 Expected frequency less than 5

If the expected frequency in one or more cells has a value of less than 5, then the result of the χ^2 test becomes unreliable. In this case the test should not proceed.

One way of dealing with this problem is to amalgamate some of the cells so that the 'combined' expected frequency is equal to 5 or more. Example 14.3 shows how this can be achieved in a simple example.

If the total number of 'frequencies', or the number of categories, is so small that this cannot be done, then it is probably necessary to repeat an enlarged experiment and record more data values.

Example 14.3

A student investigates whether there is a relationship between the colour of a car and the speed at which it is driven. Selecting an open stretch of road, he counts the numbers of cars of each colour that could be considered to be driving Very Fast, Fast, at a Moderate speed, or Slow. The observed frequencies are given in the table below:

	Red	Green
Very Fast	5	1
Fast	20	24
Moderate	12	18
Slow	3	17

What conclusion does the student reach?

The analysis is performed in the following text.

The expected frequencies for Example 14.3 are calculated as in the following table:

	Red	Green
Very Fast	2.4	3.6
Fast	17.6	26.4
Moderate	12	18
Slow	8	12

which then give a p-value for a χ^2 contingency test of $p = 0.033$.

Although the p-value might suggest that the student should accept the proposed hypothesis at $\alpha = 0.05$, it should be noted that there are two cells (in the Very Fast category) that have expected frequencies of less than 5, and the conclusion arising from the χ^2 contingency test would be unreliable.

One way forward is for the student is to combine the Very Fast data and the Fast data into just one 'Fast' category for both observed and expected frequencies:

	Observed		Expected	
	Red	Green	Red	Green
'Fast'	25	25	20	30
Moderate	12	18	12	18
Slow	3	17	8	12

The expected frequencies are now all above 5, and the χ^2 contingency test for the reduced categories now gives a p-value of 0.062. The student should therefore not accept the proposed hypothesis (at $\alpha = 0.05$).

14.1.5 The 2 × 2 contingency table (Yates correction)

A contingency table with two rows ($r = 2$) and two columns ($c = 2$) is a special case of a general $r \times c$ table.

The degrees of freedom for the 2 × 2 contingency table, $df = (2-1) \times (2-1) = 1$. However, for χ^2 calculations with $df = 1$, the normal equation for the χ^2 statistic (14.1) tends to *overestimate* the true χ^2-value. This means that the test is more likely to accept the proposed hypothesis in borderline cases when the null hypothesis is actually true – a Type I error.

It is still possible to use a χ^2 test, but when $df = 1$ it is necessary to use a revised form of the test statistic, called the Yates correction:

$$\chi^2 = \sum_i \frac{(|O_i - E_i| - 0.5)^2}{E_i} \qquad [14.3]$$

Note that $|O_i - E_i|$ means take the positive value of $O_i - E_i$ (see 3.3.3).
After the calculation of the χ^2 statistic, the rest of the chi-squared test is performed normally.

Q14.5

A student on an education course decided to investigate whether there was a significant difference between the proportions of male and female students on two courses, A and B.

The numbers are given in the table below:

	Course A	Course B
Male	32	28
Female	87	40

(i) Calculate the numbers that would be expected if there were no preference between the courses for male and female students.
(ii) Perform a contingency test, using the Yates correction, to assess whether there is a significant difference in the proportions between the courses.
(iii) Repeat the test in (ii) without using the Yates correction. Does this give a different result? If yes, which is the *best* answer?

The 2 × 2 contingency test is effectively comparing *two proportions*, and we will see in 14.3.5 that, if data analysis software is available, it would be more appropriate to use Fisher's exact test.

14.2   Goodness of Fit

14.2.1   Introduction

A *goodness of fit test* tests whether an observed distribution of occurrences in specific categories is significantly different from the distribution that would have been expected. For example, we could test whether the numbers of students obtaining various grade categories is significantly different between years.

We will see that the χ^2 test can be used to compare the *observed frequency*, O, in each category with the *expected frequency*, E. In some examples, it may be necessary to calculate expected frequencies on the basis of an expected set of *ratios*.

The procedure can also be used to test whether the distribution of some experimentally observed frequencies might follow a known distribution pattern, e.g. a normal distribution or a Poisson distribution − see the Website.

It should be noted that there are also alternative methods available in statistics software for testing for goodness of fit to a distribution, e.g. for testing whether a given set of sample data is likely to have been drawn from a normal distribution.

14.2.2   Comparison with defined ratios

Example 14.4 illustrates the use of the χ^2 statistic when comparing observed frequencies with a set of ratios that is predicted, either by theory or by historical measurements.

Example 14.4

Measurement of genotypes is expected to show types AB, Ab, aB, ab occurring in the ratio 9:3:3:1. A sample of $T = 200$ observations gave observed frequencies, O_i, of 131, 28, 32 and 9 respectively in the four groups.

Test whether this distribution of values shows a significant difference from the expected ratios.

The analysis is performed in the following text.

The relevant hypotheses for the test in Example 14.4 would be:

Proposed hypothesis, H_1: The differences between observed frequencies and the given ratios have not occurred just by chance, and hence an additional factor is significant.

Null hypothesis, H_0: The differences between observed frequencies and the given ratios might have occurred just by chance.

We set out the calculations in the form of a table (Table 14.5) in which we calculate the totals, T and R_T, for both the observed numbers and the ratio values respectively:

We can see that the *fraction* of observations falling into the category AB would be equal to $R_{AB} = 9$ of a total of $R_T = 16$. Thus the *expected* frequency for category AB would be found

Table 14.5. Goodness of fit calculation.

Category i	AB	Ab	aB	ab	Totals
O_i	131	28	32	9	$T = 200$
Ratio, R_i	9	3	3	1	$R_T = 16$
E_i	112.5	37.5	37.5	12.5	200
$(O_i - E_i)^2/E_i$	3.042	2.407	0.807	0.980	**7.236**

by multiplying the total number of observations, 200, by the fraction 9/16:

$$E_{AB} = T \times \frac{R_{AB}}{R_T} \Rightarrow 200 \times \frac{9}{16} \Rightarrow 112.5$$

In general, the *expected* frequency, E_i, for each category, i, is calculated using the simple equation:

$$E_i = T \times \frac{R_i}{R_T} \qquad [14.4]$$

The calculation of the χ^2 statistic uses equation (14.1) giving the value:

$$\chi^2 = \frac{(131 - 112.5)^2}{112.5} + \frac{(28 - 37.5)^2}{37.5} + \frac{(32 - 37.5)^2}{37.5} + \frac{(9 - 12.5)^2}{12.5} = 7.236$$

In a goodness of fit test with n data values, the number of degrees of freedom is given by $df = n - 1$. Assuming a significance level of $\alpha = 0.05$, we have from Appendix III:

$$\chi^2_{CRIT} = 7.81 (\text{for } \alpha = 0.05 \text{ and } df = 4 - 1 = 3)$$

In this case, because $\chi^2 < \chi^2_{CRIT}$, we should not accept the proposed hypothesis (at 0.05).

There is not enough evidence to claim that the apparent differences in numbers for each category occurred by chance.

The calculation for Example 14.4 is performed on the Website using both Excel and Minitab and returns a *p*-value equal to 0.065. This agrees ($p > 0.05$) with the conclusion that we should not accept the proposed hypothesis.

Q14.6

A normal playing die is thrown 120 times and we record the number of occurrences of each of the six possible scores, giving *observed* frequencies, O_i. Each of the six scores represents a different category, i.

We might *expect* that, on average, each particular score should occur 120/6 = 20 times.

The results, for both observed and expected frequencies, are shown in the table below. A quick visual examination of the data might suggest that the die is biased because, for the score '6', there appears to be a large difference between *observed* (31) and *expected* (20) frequencies.

Category, i	Score	1	2	3	4	5	6	Sum
Observed frequency	O	15	22	20	14	18	31	120
Expected frequency	E	20	20	20	20	20	20	120

Can we deduce from the data, with a confidence level of 95 %, that the die is indeed biased?

Q14.7

In a telephone switchboard for a large company, the calls should be directed randomly to the various operators. In a 5-hour period, four operators each receive the following number of calls: 44, 32, 56, 28.

Is there evidence to suggest that the calls are not actually being directed randomly?

Q14.8

A self-pollinating pink flower is expected to produce red, pink and white progeny with the following relative frequencies: 1:2:1.

A student grows 50 flowers and finds the following division of colours: Red, 18; Pink, 22; and White, 10.

Can the student conclude that his observed distribution of colours is significantly different (at 95 % confidence) from the expected frequencies?

14.2.3 Testing a simple proportion: Yates correction for df = 1

When testing just two frequencies, $n = 2$, the degrees of freedom $df = 1$, and we saw in 14.1.5 that it is then necessary to use the Yates corrected equation (14.3).

After the calculation of the χ^2 statistic using the Yates correction, the rest of the chi-squared test is performed normally.

Example 14.5

A random sample of 50 frogs is taken from a lake, and it is found that 37 are female and 13 are male. The expected proportion, Π_O, of female to male frogs is 60:40.

Are the observed numbers significantly different from those expected?

Use both equations (14.1) and (14.3) to calculate the value for χ^2_{STAT}.

Observed	O_i	37	13	
Expected	E_i	30	20	
	$O_i - E_i$	7	−7	
Normal χ^2	$(O_i - E_i)^2/E_i$	1.63	2.45	$\chi^2 = 1.63 + 2.45 = 4.08$
Yates χ^2	$(|O_i - E_i| - 0.5)^2/E_i$	1.41	2.11	$\chi^2 = 1.41 + 2.11 = 3.52$

The critical value for $df = n - 1 = 2 - 1 = 1$ and $\alpha = 0.05$ is:

$$\chi^2_{CRIT} = 3.84$$

We find that, using the normal χ^2 value (14.1), $\chi^2_{STAT} > \chi^2_{CRIT}$, suggesting that we should accept the proposed hypothesis. However, using the Yates correction (14.3), $\chi^2_{STAT} < \chi^2_{CRIT}$, suggesting that we should not accept the proposed hypothesis.

The Yates correction gives a more cautious conclusion.

The above test effectively compares a *single proportion*, P, to a specific value, Π_O, and we will see in 14.3.5 that, if data analysis software is available, it would be more appropriate to use a Fisher's Exact Test.

14.3 Tests for Proportion

14.3.1 Introduction

When frequency counts fall into one of two categories (e.g. yes or no, male or female, wet or dry) it is possible to express the ratio as a simple proportion. If, in a questionnaire, 120 people respond Yes to a specific question and 80 respond No, then the *proportion* of Yes responses is:

$$P = \frac{120}{120 + 80} \Rightarrow \frac{120}{200} \Rightarrow 0.60 \Rightarrow 60\%$$

In this unit we will see that the fundamental statistics of simple proportion are based on the binomial distribution, which leads to Fisher's exact test. However, the χ^2 test can also be used with the Yates correction and, under suitable conditions, the normal distribution can also be used to calculate a confidence interval for the true proportion.

The one-proportion Fisher's exact test looks for a difference between the measured proportion, P, of responses in one sample and a specific value for the proportion, Π_O.

The two-proportion Fisher's exact test looks for a difference between the measured proportions, P_A and P_B, of two separate samples.

It is important to note that the value of the proportion is not itself sufficient data for the test. It is essential to have the actual *frequencies of events* which fall into the two categories of the proportion.

14.3.2 Exact one-proportion test

A simple proportion, where the outcome is either Y or N $(= \overline{Y})$, is a direct binary choice, and the associated probabilities are determined by the statistics of the binomial distribution (8.4).

The statistics of an exact test using the binomial distribution were developed in Example 9.4. These statistics form the basis of Fisher's exact test which would normally be used in software analysis.

Example 14.6

It is expected that, for a particular species of frog, 60 % will be females and 40 % males.

A random sample of 50 of these frogs is taken from a lake, and it is found that 37 are female and 13 are male. Is this proportion significantly greater than expected? (See also Example 14.5.)

The analysis is performed in the following text.

In Example 14.6 the hypotheses will be:

Proposed hypothesis, H_1: The proportion of female frogs is greater than 0.6.
Null hypothesis, H_0: The proportion of female frogs is equal to 0.6.

If the null hypothesis were true, then the probability that any given frog would be female is 0.6, and we can use the binomial theorem to calculate the probability of recording different numbers of female frogs.

The shaded areas in Figure 14.1 show that the probability of recording 37 or more females is 0.0280. This is the p-value for this test, and since $p < 0.05$, we would accept the proposed hypothesis for $\alpha = 0.05$.

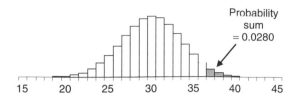

Figure 14.1 Probabilities for specific numbers of female frogs for Example 14.6.

The calculation for Example 14.6 is performed on the Website using Minitab for a 1-tailed, one-proportion, Fisher's exact test, returning the same p-value equal to 0.028.

Example 14.7

Using the same data as in Example 14.6, test whether the observed proportion of 37 females out of 50 is *different* from the expected 0.6 ratio.

The worked answer is given in the text.

Example 14.7 is now a 2-tailed problem. On careful inspection of Figure 14.1, it can be seen that the distribution is not exactly symmetrical, and we cannot expect that the 2-tailed p-value is double the one-tailed p-value as would be expected from equation (9.1) for a *symmetrical* distribution.

The calculation for Example 14.7 is performed on the Website using Minitab for a 2-tailed, one-proportion, Fisher's exact test, returning a p-value equal to 0.044. At a significance level of 0.05 we would accept the proposed hypothesis.

Q14.9

A coin is tossed 100 times and records 60 heads.

Refer to the calculations performed for Examples 9.4 and 9.5 to find the p-values that would be returned by:

   (i) a 1-tailed, one-proportion test for possible bias; and
  (ii) a 2-tailed, one-proportion test for possible bias.
 (iii) Explain why, in this case, you *would* expect that the answer for (ii) is double the answer for (i), although the same is not true for Examples 14.6 and 14.7.

14.3.3 Confidence interval (approximate) of the true proportion

The unknown *true* proportion for a system is a *parameter* of the system and we will describe it by the Greek symbol Π (capital pi). The true value, Π, is the proportion of Y outcomes that would be achieved if the whole *population* of the system were measured.

The experimentally measured proportion, P, from a *sample* of n 'trials' is the *best estimate* for both the unknown true proportion, Π, of the population, and also for the probability, p, of outcome Y for each individual event (8.4.2).

Provided that the total number of 'trials', n, recorded is sufficiently large and that the probability, p, for each event is not very close to either 0.0 or 1.0 (i.e. $np(1 - p) \geqslant 5$, see 8.4.7), then it is possible to use the **normal distribution** as a *good approximation* for the binomial distribution.

Using the normal approximation we can calculate the 95 % confidence interval (8.3.2) for the value of the true proportion, Π:

$$CI_{95\%} \approx P \pm 1.96 \times \sqrt{\frac{P \times (1 - P)}{n}} \qquad [14.5]$$

Example 14.8

Take the same data as in Example 14.6, with 37 female frogs out of 50, and use the normal approximation to the binomial distribution to calculate the 95 % confidence interval for the true proportion of females in the population of frogs.

The experimental data gives:

Proportion, $P = 37/50 = 0.74$
Total number, $n = 50$

Hence, using equation (14.5):

$$CI_{95\%} \approx 0.74 \pm 1.96 \times \sqrt{\frac{0.74 \times (1 - 0.74)}{50}} \Rightarrow 0.74 \pm 0.12$$

which gives a 95 % confidence interval, $CI_{95\%}$, from 0.62 to 0.86.

It is important to note that the confidence interval in Example 14.8 assumes the normal approximation to the binomial distribution.

Q14.10

A coin is tossed 100 times and records 60 heads.

(i) Use the normal distribution approximation to calculate the 95 % confidence interval for the true probability of recording a 'head' in a single toss.
(ii) Does your answer in (i) suggest that, at a significance level of 0.05, the coin is biased and does not have a true probability of 0.5?

14.3.4 Comparison of statistical tests

Example 14.9 illustrates the fact that it is not uncommon to find that there is more than just one way of addressing a given analytical problem.

Example 14.9

Compare Examples 14.5, 14.7 and 14.8 to investigate three ways of addressing the same two-tailed, one-proportion test.

Do the conclusions differ on whether the proportion of 37 out of 50 is significantly different from 0.6?

Chi-squared test

The standard calculation of the chi-squared test (Example 14.5) identifies a significant difference, but the Yates correction is more cautious, suggesting that there is no significant difference.

Confidence interval calculation

The confidence interval (Example 14.8) for the true proportion gives a range from 0.62 to 0.86, and as 0.6 is just outside this range, this suggests a significant difference.

One-proportion exact test

The p-value of 0.044 from Fisher's exact test (Example 14.7) is just in favour of a significant difference.

Clearly from Example 14.9, the approximations using the chi-squared distribution or the normal distribution can return slightly different values and conclusions. However, provided the relevant software is available, Fisher's exact test is easy to perform and gives an 'exact' result based on the binomial distribution.

14.3.5   Exact two-proportion test

A two-proportion test looks for a difference between two experimental proportions, P_A and P_B.

Example 14.10

A student on an education course decided to investigate whether there was a significant difference between the proportions of male and female students on two courses, A and B.

The numbers are given in the table below.

	Course A	Course B
Male	32	28
Female	87	40

The analysis is performed in the following text.

In Example 14.10 the proportions for male students are:

$$\text{Number in course A}, n_A = 119, \text{giving a proportion of males}: P_A = \frac{32}{119} \Rightarrow 0.269$$

$$\text{Number in course B}, n_B = 68, \text{giving a proportion of males}: P_B = \frac{28}{68} \Rightarrow 0.412$$

The calculation is performed on the Website using Minitab for a 2-tailed, two-proportion, Fisher's exact test, returning a p-value equal to 0.051. At a significance level of 0.05 we would not accept the proposed hypothesis that the two proportions, P_A and P_B, were different.

It is important to note that, when testing for proportion, it is still necessary to know values for the numbers n_A and n_B, as well as the actual proportions P_A and P_B.

Q14.11

Compare the analysis used for Example 14.10 with the χ^2 test used in Q14.5 for the same problem. Do they both reach the same conclusion?

15

Experimental Design

Overview

There is an enormous variety of possible ways to design experiments, and it is important that each design is tailored to fit the particular investigation. Consequently, it is not possible to give a step-by-step guide to producing the correct design in any given situation. The only practical approach is to develop each new experiment on the basis of knowledge and experience of the range of fundamental design principles and data analysis techniques that may be appropriate. This chapter aims to provide an overview of key aspects of experimental design and their links with suitable statistical tests.

The objective of good experimental design is to reduce the errors in any measurement and improve the confidence in the final experimental result. In particular, a good experimental design will counteract the effects of possible systematic errors, and take account of the magnitude of possible random errors.

The process of experimental design proceeds hand in hand with the availability of statistical techniques for analysing data. In particular, readers are recommended to read Chapters 9 to 14 in conjunction with this chapter, and follow the development of statistical tests in parallel with their use in supporting good experimental design.

15.1 Principal Techniques

15.1.1 Introduction

In general there are five main techniques used in experimental design:

Replication – counteracts the effect of random experimental uncertainties, and makes their effect more explicit in the analysis.
Randomization – counteracts hidden systematic errors (bias) and makes the effects of random subject uncertainties more explicit in the analysis.
Repeated measures or pairing – counteracts for systematic differences between subjects by linking subject responses in all factor levels.

Essential Mathematics and Statistics for Science 2nd Edition   Graham Currell and Antony Dowman
Copyright © 2009 John Wiley & Sons, Ltd

Blocking of factors – used to separate the effects of known factors, including possible systematic (bias) effects.

Allocation of subjects – used to counter the effects of remaining unknown factors and bias.

15.1.2 Replication

Replication is the process of repeating the experimental measurement under exactly the same conditions. With replicated measurements, the only variations are those inherent in the experimental process itself – all other factors remain the same.

If the replication is performed with the same subject – *within subjects* – then any variation will only be due to *measurement* uncertainty, e.g. repeating the measurement of blood–alcohol level on the *same sample* of blood will demonstrate the variability of the measurement itself.

If the replicate experiments involve different subjects – *between subjects* – then any variation will be due to *subject* uncertainty in addition to *measurement* uncertainty, e.g. the growth of several bacteria colonies exposed to the *same* experimental conditions will demonstrate the variability of bacterial growth.

The process of replication provides information that can be used directly to calculate the magnitude of the experimental uncertainty. Example 15.1 illustrates the calculation of the *confidence interval* (8.2.4).

Example 15.1

An experimental measurement of the sodium content of a water sample is repeated, giving five replicate results:

Sodium content (mg L^{-1})	31.2	32.3	31.9	31.7	31.5

Give the best estimate for the true sodium content, together with an estimate of measurement uncertainty and the 95 % confidence interval.

Mean value, $\bar{x} = 31.72$

Sample standard deviation, $s = 0.4147$

Standard uncertainty [8.4], $u_x = s/\sqrt{n} = 0.415/\sqrt{5} = 0.185$

The t-value (Table 8.2) for $df (= n - 1 = 4)$ and default significance level $\alpha (= 0.05)$, $t = 2.78$

Confidence deviation [8.6], $Cd_{95\%} = t \times u_x = 2.78 \times 0.185 = 0.515$

Confidence interval [8.5] of the mean value, $CI_{95\%} = \bar{x} \pm Cd_{95\%} = 31.72 \pm 0.52$

The process of *linear regression* (13.2.2) is also based on a form of replication. Two points are all that is required to define a straight line, so that additional points provide 'replicate' information for the calculation of the 'best-fit' straight line. The calculation of experimental uncertainty is based on the spread of points (*residuals*) around the 'best-fit' straight line, and is expressed using the standard error of regression [13.4].

15.1.3 Randomization

Randomization is the process of making a truly random choice in the allocation of the subjects to the different factor levels.

In an experiment where the major uncertainty may be due to a subject variability or bias, it is necessary to make a random selection of subjects between samples or factor levels. Although we may believe that we could pick subjects randomly (e.g. seedlings for a growth experiment, patients for a drug trial), experience shows that we would almost always be biased subconsciously in some way.

The process of random allocation can be performed simply by giving each subject a number, and then allocating each subject to a factor level (treatment) by reference to a set of random numbers.

Example 15.2

As a simple example of random allocation, use the following set of random numbers:

$$7, 3, 7, 8, 2, 6, 6, 7, 9, 1, 2, 2, 9, 4, 6, 6, 2, 6, 5, 3, 8, 2, 4$$

to produce a random allocation of the eight letters ABCDEFGH to two samples, X and Y, of four letters each.

Allocate a number to each of the letters:

$$A = 1, B = 2, C = 3, D = 4, E = 5, F = 6, G = 7, H = 8$$

Work through the list of random numbers and allocate the letter that corresponds to the first number to sample X, and the letter that corresponds to the next number to the sample Y. Continue with allocations to alternative samples. If a number appears which does not correspond to an available letter just pass on to the next number.

Resulting allocation:

Sample X: G, H, F, D

Sample Y: C, B, A, E

Drugs trials are conducted as 'double blind' trials, in which the patients are allocated randomly to the drug or placebo option in such a way that:

- the medical staff do not know which patients are taking the drug, so that their treatment does not become biased; and
- the patients do not know whether they are taking the drug, so that their response to the treatment is not affected subconsciously.

An experiment where subjects are randomly allocated to the different levels of a factor is often called a **completely randomized design**. The experiment described in Example 11.1 is an example of such a design.

15.1.4 Blocking of factors

Blocking is a process in which groups (blocks) of subjects are identified with particular combinations of levels for different factors. The process of 'blocking' is aimed at separating the effects of known, and possibly unknown, factors that may affect the outcome of the experiment.

The analysis performed in Example 11.2 assumed that it was a *completely randomized design* experiment where the effects of three different catalysts were each measured with four replicates. A one-way ANOVA gave a *p*-value of 0.0845, which suggested that, at a significance level of $\alpha = 0.05$, there was *not* sufficient evidence to identify any difference between the catalysts.

However, using the same data, Example 11.4 treated the 'days' as a possible *second factor*, and 'blocked' the repeated measurements into columns for each of the four days. It then analysed the two-factor data using a two-way ANOVA, giving *p*-values for both catalyst and day effects:

$$p(\text{catalyst}) = 0.0173$$

$$p(\text{day}) = 0.0328$$

which suggested that, at a significance level of $\alpha = 0.05$, there was now sufficient evidence for a difference between the catalysts, and also between the days.

The effect of *blocking* in Example 11.4 enabled the analytical software to allow for systematic variations between the days and provide a more powerful (9.4.5) analysis of the possible effects of the catalysts.

The experimental design used in Example 11.4 is a **randomized block design**.

It is possible to extract still more information from the basic two-factor *randomized block design* if we also add *replication*; that is, for every 'cell' in the table, we make at least two replicate measurements. In this case, the replicate information allows an ANOVA calculation to estimate the magnitude of the experimental variation, and then any remaining variation in the data can be explained by an **interaction** (11.3) between the two factors.

15.1.5 Repeated measures or pairing

An effective way of counteracting *subject* variation is to ensure that each subject is 'tested' at all levels of each factor, and that the results are compared individually for each subject. The simplest example is the paired *t*-test (10.3) where the differences of response, d_i, are calculated between the responses for the two factor levels, and then a one-sample test applied to the set of difference values. See Example 10.7 which uses a *pairing* of data values to reduce the uncertainty due to variations between subjects.

The procedure of testing the same subject at all levels of a test factor is also used in more complex experiments, and it is then referred to as a **repeated measures** experiment.

Within 'repeated measures' or 'paired' designs it is necessary to remember that a subject might gain 'experience' from one test which can then affect a subsequent test; for example, in a sports test there may be an element of training in having already performed the test under one set of conditions. Example 15.3 shows that this problem can be counteracted by using a *cross-over* design, where different groups of subjects are allocated to do the repeated measures in a reverse order.

Example 15.3

As part of a project to investigate the factors that affect reaction times, 10 subjects perform standard reaction tests whose timing is measured through a computer.

Five subjects (1 to 5) perform the test, first with a high level of background noise, and then repeat the test with no background noise. Five other subjects (6 to 10) perform the same tests, but first with no noise, followed by the test with the high level of background noise.

The experiment is a 'cross-over' design to ensure that the *order of the testing* is opposite for the two groups. This counteracts the possibility that performing one test may affect performance in the second test, e.g. through 'training'.

Perform a paired t-test to investigate whether there is any significant difference (at 0.05) between the reaction times under the two conditions. The results are given in the table below, where Difference = Noise − No noise.

Subject	1	2	3	4	5	6	7	8	9	10
Noise	6.5	8.2	4.0	7.5	4.8					
No noise	5.6	8.7	3.4	6.2	4.4	6.9	4.2	5.9	3.3	7.6
Noise						7.6	4.0	6.3	4.3	8.0
Difference	0.9	−0.5	0.6	1.3	0.4	0.7	−0.2	0.4	1.0	0.4

Each subject links unique 'pairs' of data in the 'No noise' and 'Noise' data sets.

Take the differences, d_i, between the values in each data pair.

Sample standard deviation of the differences, $s = 0.5395$

Mean value of differences, $\bar{x} = 0.50$

Using the one-sample t_{STAT} in [10.1], compare $\bar{x}$ with $\mu_O = 0$:

$$t_{STAT} = \frac{(\bar{x} - \mu_O)}{s/\sqrt{n}} = \frac{0.50}{0.5395/\sqrt{10}} = 2.93$$

Degrees of freedom, $df = n - 1 = 9$

For the two-tailed test at 0.05 significance, $t_{CRIT} = 2.26$

Since $t_{STAT} > t_{CRIT}$, we accept the proposed hypothesis that there is a difference in reaction times between the different conditions.

15.1.6   Allocation of subjects

We saw in 15.1.3 that, where there are *no systematic differences* between the subjects, they should be allocated *randomly* to the various factor levels.

However, in situations where there may be bias between subjects or between the measurement of the subjects, it is appropriate to allocate the subjects in such a way that is likely to counteract the possible effects of the bias.

In particular, hidden bias between subjects might give the appearance of a significant factor effect when, in fact, no such effect actually exists. For example, in testing for difference in athletic performance, an allocation of *young* athletes to one sample, and *older* athletes to the other, might produce a biased effect due to age that masks the factor being tested. Undetected bias can lead directly to a Type I error (9.4.4).

We can illustrate this situation with the following examples that investigate the effect of subject allocation.

Example 15.4

An analytical laboratory decided to test whether there was any difference between the effects of three different catalysts ($C1, C2, C3$) in an industrial process. The process was repeated using three different temperatures ($T1, T2, T3$) for each catalyst.

The yields from each experiment, recorded in the table below, are those that would be recorded assuming that *no systematic measurement errors* are made in recording the data (see also the following Example 15.5).

	$T1$	$T2$	$T3$
$C1$	39	33	32
$C2$	36	36	31
$C3$	32	30	29

Use a two-way ANOVA to test whether there is a significant difference between the effects of the three catalysts.

The analysis is performed in the following text.

The results of a two-way ANOVA for Example 15.4 give:

Table 15.1. ANOVA results for Example 15.4.

Variation	SS	df	MS	F	P-value	F crit
Catalyst	34.88889	2	17.44444	5.607143	0.069122	6.944276
Temperature	37.55556	2	18.77778	6.035714	0.061946	6.944276
Error	12.44444	4	3.111111			
Total	84.88889	8				

The results in Table 15.1:

p-value for catalyst $= 0.069$

p-value for temperature $= 0.062$

show that there is not sufficient evidence (at 0.05) to suggest that either the catalyst or temperature has a significant effect on the experiment yield.

The next example considers the effect of possible bias due to a specific allocation of 'subjects'. In this case each 'subject' represents an experiment carried out by a particular scientist.

Example 15.5

Consider the situation where the experiments given in the previous example, Example 15.4, are actually carried out by three scientists, A, B and C, two of whom produce systematic errors:

Scientist A produces results that are too high by 1.0.
Scientist B produces accurate results.
Scientist C produces results that are too low by 1.0.

Scientist A measures the performance of Catalyst 1 at each of the three temperatures, and Scientists B and C measure the performances of Catalysts 2 and 3 respectively – the allocation of 'subjects' to scientists is shown in Table 15.2(a).

Table 15.2. (a) Allocation of scientists. (b) Experimental results.

	T1	T2	T3			T1	T2	T3
C1	A	A	A	leads	C1	40	34	33
C2	B	B	B	to data values:	C2	36	36	31
C3	C	C	C		C3	31	29	28
	(a)					(b)		

The actual data values recorded by the scientists can then be calculated from the data in Example 15.4, and these values are given in Table 15.2(b).

Use a two-way ANOVA to decide what conclusions would be drawn from these results.

The analysis is performed in the following text.

The results of a two-way ANOVA for Example 15.5 give the results:

$$p\text{-value for catalyst} = 0.025$$

$$p\text{-value for temperature} = 0.062$$

which *appear to show* that the catalyst *does* have a significant effect. We know from the analysis in Example 15.4 that this is a *false* result (Type I error). It has arisen because of the *bias between the different scientists*.

Example 15.6 now investigates the possible effects of different subject allocation patterns.

Example 15.6

Repeat the calculations performed in the previous example, Example 15.5, but for the different allocations for the three scientists as given below:

	$T1$	$T2$	$T3$
$C1$	A	B	C
$C2$	A	B	C
$C3$	A	B	C

(1)

	$T1$	$T2$	$T3$
$C1$	B	A	A
$C2$	B	C	A
$C3$	C	C	B

(2)

	$T1$	$T2$	$T3$
$C1$	A	B	C
$C2$	C	A	B
$C3$	B	C	A

(3)

The analysis is performed in the following text.

The results of a two-way ANOVA carried out for each of the allocations in Example 15.6 give the results:

p-value for catalyst =	0.069	0.013	0.217
p-value for temperature =	0.021	0.049	0.200
Subject allocation:	(1)	(2)	(3)

Compared with the 'true' results of Example 15.4, it can be seen that subject allocation (1) gives a false result for the effect of temperature, and allocation (2) gives false results for both the effects of the catalyst and temperature.

In subject allocations (1) and (2) the systematic errors in measurement have created the *appearance* of significant effects in the results, leading to Type I errors. However, the results for subject allocation (3) show *increased p*-values (compared with Example 15.4), which reduces the probability of making a Type I error. This is consistent with the fact that greater errors have been introduced into the measurement process because of the bias of two scientists.

The important aspect of allocation (3) is that *each of the values of A, B and C appears once only in each row and column*. By ensuring that every level of every factor has at least one occurrence of each subject type (A, B or C), the differences between subject types are more likely to be reduced.

Any arrangement of items where each item appears just once in every row and column is called a **Latin square**. A very popular example of a Latin square is the number arrangement for the Japanese Sudoku puzzle. Tables of Latin squares are available for each size of square, and the particular one to be used should be selected at random.

In the above example, the allocation of scientists to the labels A, B and C should be made at random.

As a final note, it is important to understand that any particular allocation of subjects may *coincidentally* work with other variations in measurement to produce *increased* errors. However, provided that the proper procedures are followed, then it is *more likely* that an appropriate allocation of subjects will *reduce* the probability of errors.

15.2 Planning a Research Project

Carrying out a science project as an undergraduate is likely to be the first major piece of original work carried out by a student.

Many student projects are marred by the fact that the student did not anticipate how the data would be analysed and failed to collect data that was essential for the process of analysis. Too often, a student reports 'no measurable effect was discovered' because the design of the experiment did not ensure that the data collected would match the methods used to analyse it.

If the process of data collection is not planned to match the requirements of the analytical process, then information will be lost. A loss of information will then lead to:

- increased uncertainty in the experimental results; and
- a reduction in the 'power' of a hypothesis test (9.4.5).

Before starting the project, the student should:

- be clear about the type of investigation – hypothesis testing or direct measurement;
- focus clearly on the types of quantitative conclusions that could be drawn from the results of the experiments;
- identify the possible sources of uncertainty (random and systematic) in the measurements;
- decide what mathematical or statistical process could be used for analysing the data;
- create an experiment design;
- plan the complete experimental process from measurement to analysis
- carry out a short 'pilot' experiment to assess the methods to be used, and to obtain some example data; and

- check all stages of the analytical process using 'test' data derived either from the pilot experiment, or by taking an intelligent guess at the sort of values that might be produced in the experiments.

A set of short 'pilot' measurements at the very beginning of the project can be extremely useful in picking up any unforeseen problems at an early stage, and can often suggest useful modifications to the main experimental plan. In addition, 'pilot' measurements will produce 'test' data with appropriate precision and variance suitable for checking the data analysis procedures.

This procedure of working through the whole process of data collection and analysis allows the student to check that nothing has been left out, and prevents the embarrassing scenario of preparing to write up the project only to realize that an essential element of data was never recorded!

Appendix I: Microsoft Excel

A spreadsheet software package (e.g. Excel) is very convenient for managing, recording and displaying scientific data. It can also be used directly for many of the calculations involved in scientific data analysis. In addition, where it is necessary to use dedicated packages to perform additional analysis, the spreadsheet can be used to prepare the data in a form which is then easy to copy and paste into the new package.

Calculations in Excel can be performed by entering equations directly into the spreadsheet cells, using normal mathematical operations: $+$ 'add', $-$ 'subtract', $*$ 'multiply', $/$ 'divide', and $\char`^$ for to the 'power of'. Excel also provides Functions, which can be used to perform more complex calculations, Data Analysis Tools which perform specific statistical analyses, and Solver which can provide a numerical solution to complex problems. The Trendline option is also very useful for drawing best-fit lines of regression in $x-y$ graphs (4.2.2).

Tutorials on the use of Excel for scientific data analysis are provided on the Website.

Functions

Excel uses many different **functions** to perform specific calculations. A typical example is the TTEST function for a two-sample t-test (10.2.1):

$$=\text{TTEST}(array1, array2, tails, type)$$

The expression is placed in an Excel cell. The four components inside the brackets are called the **arguments**, and identify the values that the TTEST function will use in its calculation.

The values for the arguments of functions can be entered:

- directly as numbers, e.g. '=SQRT(16)' calculates the square root of 16, giving the answer 4;
- as cell references, e.g. '=SQRT(C8)' gives the square root of the number that is found in C8;
- as arrays of data, e.g. '=SUM(B2:E4)' would calculate the sum of all the values in the rectangular block of cells defined by B2 in the top left-hand corner and E4 in the bottom right-hand corner;
- as calculations using equations and/or other functions, e.g. '=SQRT(2*SUM(A4:B5))' would calculate the square root of twice the sum of the four values held in the block of cells between A4 and B5.

For example, the entry '=TTEST(A5:A10,B5:B9,1,2)' in cell D2 would perform a 1-tailed, two-sample t-test between one sample of six data values held in cells A5 to A10 and a second sample of five data values held in cells B5 to B9, leaving the calculated p-value in cell D2.

A list of some of the most commonly used functions for scientific data analysis are given at the end of this appendix. There are also some special *Array Functions* in Excel, which are sometimes difficult to use. However, two useful array functions, LINEST and FREQUENCY, are included in the list. Further information on their use is given on the Website.

Data Analysis Tools

The **Data Analysis Tools** facility contains a range of specific statistical calculations, e.g. *ANOVA tests, F-Test, t-Tests* and *Regression*. Using Data Analysis Tools is slightly different in Excel 2003 and Excel 2007.

In Excel 2003, these are available from the Menu bar using Tools > Data Analysis. If this option does not appear, it is necessary to go to Tools > Add-Ins, select the Analysis ToolPak and then click OK. This should install the Data Analysis tools. When next selecting Tools from the Menu, the option should be available, otherwise it may require the Excel installation disc or assistance from a network manager.

In Excel 2007, Data Analysis may well need to be added in. The procedure is to click the 'Office Button' icon at the top left of the Excel window, click 'Excel Options' at the bottom, then 'Add-Ins', Click 'Analysis ToolPak', then 'Go. . .', and finally 'OK'. Having completed the add-in process, clicking 'Data' in the top menu will open an 'Analysis' section with 'Data Analysis' within it. Click 'Data Analysis' and the analysis tools will appear.

Note that the results of the calculations using Data Analysis Tools are *not dynamic*; if the original data is changed it is necessary to *recalculate* the results. This is in contrast to all Excel functions, which are *dynamic* and automatically change immediately the original data is changed.

Solver

Solver is an interactive programme within Excel that can be used to provide 'solutions' to a variety of mathematical problems. It is first necessary to define the mathematical conditions that must be met (e.g. two equations must be simultaneously true – see 3.5.3). Solver then uses a numerical process of iteration to find values for the unknown variables that will make those conditions as true *as required*. Note that the solution is not absolutely exact, but for the majority of simple problems the errors are extremely small, and it is possible to define how accurate the iterative process should be before it stops to present the results.

The use of Solver is introduced on the Website.

Solver may well need to be added in, by finding 'Solver' within the 'Add-Ins' in a similar way to Data Analysis Tools above.

Trendline for 'Line of Best Fit' in x–y Graphs

The commands for using a Trendline are again slightly different in Excel 2003 and Excel 2007.

In Excel 2003, highlighting a chart gives a Chart menu along the top, from which 'Add Trendline' should be clicked. Under the Type tab the default Linear Trend/Regression type should be selected by entering 'OK'. Under the Options tab it is possible to display the equation on the chart, set the intercept to zero if required, and forecast (extend) the Trendline forwards and backwards.

In Excel 2007, highlighting the chart gives the Layout menu with a Trendline section. By clicking on Trendline it is then possible to choose 'Linear Trendline' or 'More Trendline Options' at the bottom and then choose Type (e.g. Linear), Display Equation on chart, etc., as for Excel 2003 above.

The quicker way in both versions of Excel is to click the right mouse button on any data point in the chart to obtain the 'Add Trendline' option.

Useful Functions for Scientific Data Analysis

Angles (2.4.9)

SIN(θ), **COS**(θ), **TAN**(θ) calculate the respective trigonometric functions for the *angle*, θ, assuming that θ is given in radians.

ASIN(x), **ACOS**(x), **ATAN**(x) calculate the respective inverse trigonometric functions for the value of x, giving the result in radians.

DEGREES(θ) converts the *angle*, θ, in radians, to the equivalent value in degrees.

RADIANS(θ) converts the *angle*, θ, in degrees to the equivalent value in radians.

PI() with no value in the argument returns the value for the constant π.

Straight line calculations for the 'best-fit' straight line passing through data points defined by *y-data* and *x-data* (4.2.4)

SLOPE(*y-data, x-data*) calculates the slope (gradient).

INTERCEPT(*y-data, x-data*) calculates the intercept.

LINEST(*y-data, x-data*, FALSE, FALSE) calculates the slope (gradient) of the best-fit straight line that is forced to *pass through the origin* of the graph. This is an array function, but by entering 'FALSE' twice as shown, it acts as a simple function.

STEYX(*y-data, x-data*) calculates the standard error of regression (13.2.2).

Logarithms and exponential (5.1.4 and 5.1.5)

LOG(x) calculates the logarithm to base 10 of x.

$10^\wedge x$ is a simple expression used to calculate 10 to the power of x.

LN(x) calculates the natural logarithm to base e of x.

EXP(x) calculates e to the power of x.

Logical values and probability

AND(*logic1, logic2*) returns the logical value of *logic1* AND *logic2* (7.4.3).

OR(*logic1, logic2*) returns the logical value of *logic1* OR *logic2* (7.4.4).

NOT(*logic*) returns the logical value of NOT *logic* (7.4.5).

IF(*logic, ans1, ans2*) returns the value *ans1* if the statement *logic* is true, otherwise it returns the value *ans2*.

FACT(x) calculates the factorial, $x!$, of the integer x (7.5.2).

PERMUT(n, r) calculates the value of $_nP_r$ (7.5.3).

COMBIN(n, r) calculates the value of $_nC_r$ (7.5.4).

Basic statistical calculations

SUM(*data*) calculates the sum of the values in *data* (7.2.3).

AVERAGE(*data*) calculates the average of the values in *data* (7.2.4).

STDEV(*data*) calculates the *sample* standard deviation of the values in *data* (7.2.5).

STDEVP(*data*) calculates the *population* standard deviation of the values in *data* (7.2.5).

VAR(*data*) calculates the *sample* variance of the values in *data* (7.2.5).

VARP(*data*) calculates the *population* variance of the values in *data* (7.2.5).

Statistical distributions

NORMDIST$(x, \mu, \sigma, logic)$ calculates the probability at a value, x, for a normal distribution with mean μ and standard deviation σ. If *logic* is TRUE then it returns the *cumulative* probability, otherwise it gives the *probability density* (8.1.3).

BINOMDIST$(r, n, p, logic)$ calculates probability in a binomial distribution for r results from n trials, each with a probability p of occurring. If *logic* is TRUE then it returns the *cumulative* probability, otherwise it gives the *probability density* (8.4.2).

POISSON$(r, \mu, logic)$ calculates probability in a Poisson distribution for r results for an expected mean number μ. If *logic* is TRUE then it returns the *cumulative* probability, otherwise it gives the *probability density* (8.4.4).

CONFIDENCE This function should *not* be used for the confidence interval of *sample* data (8.2.4) because it assumes a *population* standard deviation.

FREQUENCY(*data_array,bin_array*) This is an array function which returns the frequency of values in the *data_array* that fall within each of the intervals described by the *bin_array*. It requires careful use – see the Website.

Statistical tests

TTEST(*array1, array2, tails, type*) calculates the p-value for a two-sample or paired t-test with the data values held in *array1* and *array2* (10.2.2 and 10.3.2).

TDIST$(t_{STAT}, df, tails)$ returns the p-value corresponding to a calculated value of t_{STAT} for degrees of freedom, df, with the given number of *tails* (10.1.3 and 10.2.3).

TINV(α, df) returns the 2-tailed value of t_{CRIT} for a significance level of α and degrees of freedom, df (10.1.3 and 10.2.3).

FTEST(*array1, array2*) calculates the p-value for a 2-tailed F-test with the data values held in *array1* and *array2* (10.4.2).

FDIST(F_{STAT}, df_N, df_D) returns the 1-tailed p-value for an F-test with degrees of freedoms df_N and df_D for the numerator and denominator. The 2-tailed value can be obtained by using $2 \times$FDIST(F_{STAT}, df_N, df_D) (10.4.3).

FINV(α, df_N, df_D) returns the 1-tailed value of F_{CRIT} for a significance level of α and degrees of freedoms df_N and df_D for the numerator and denominator. The 2-tailed value can be obtained by using FINV$(\alpha/2, df_N, df_D)$ (10.4.3).

CHITEST(*observed, expected*) calculates the p-value for a χ^2 test based on the *observed* and *expected* frequencies (14.1.2 and 14.2.2). Excel does *not* perform a Yates correction (14.1.5).

CHIDIST(χ^2_{STAT}, df) returns the p-value corresponding to a calculated value of χ^2_{STAT} for degrees of freedom df (14.1.2 and 14.2.2).

CHIINV(α, df) returns the value of χ^2_{CRIT} for a significance level of α and degrees of freedom df (14.1.2 and 14.2.2).

CORREL(*array1, array2*) and **PEARSON**(*array1,array2*) both calculate the linear correlation coefficient, r, between two samples of data held in *array1* and *array2* (13.1.2).

Appendix II: Cumulative z-areas for standard normal distribution

Area shown is the cumulative probability between $z = -$ infinity and the given value of z

Initial example for $z = 0.445$ gives Probability Area = 0.6718

z	0.00	0.01	0.02	0.03	0.04	0.05	0.06	0.07	0.08	0.09
0.0	0.5000	0.5040	0.5080	0.5120	0.5160	0.5199	0.5239	0.5279	0.5319	0.5359
0.1	0.5398	0.5438	0.5478	0.5517	0.5557	0.5596	0.5636	0.5675	0.5714	0.5753
0.2	0.5793	0.5832	0.5871	0.5910	0.5948	0.5987	0.6026	0.6064	0.6103	0.6141
0.3	0.6179	0.6217	0.6255	0.6293	0.6331	0.6368	0.6406	0.6443	0.6480	0.6517
0.4	0.6554	0.6591	0.6628	0.6664	0.6700	0.6736	0.6772	0.6808	0.6844	0.6879
0.5	0.6915	0.6950	0.6985	0.7019	0.7054	0.7088	0.7123	0.7157	0.7190	0.7224
0.6	0.7257	0.7291	0.7324	0.7357	0.7389	0.7422	0.7454	0.7486	0.7517	0.7549
0.7	0.7580	0.7611	0.7642	0.7673	0.7704	0.7734	0.7764	0.7794	0.7823	0.7852
0.8	0.7881	0.7910	0.7939	0.7967	0.7995	0.8023	0.8051	0.8078	0.8106	0.8133
0.9	0.8159	0.8186	0.8212	0.8238	0.8264	0.8289	0.8315	0.8340	0.8365	0.8389
1.0	0.8413	0.8438	0.8461	0.8485	0.8508	0.8531	0.8554	0.8577	0.8599	0.8621
1.1	0.8643	0.8665	0.8686	0.8708	0.8729	0.8749	0.8770	0.8790	0.8810	0.8830
1.2	0.8849	0.8869	0.8888	0.8907	0.8925	0.8944	0.8962	0.8980	0.8997	0.9015
1.3	0.9032	0.9049	0.9066	0.9082	0.9099	0.9115	0.9131	0.9147	0.9162	0.9177
1.4	0.9192	0.9207	0.9222	0.9236	0.9251	0.9265	0.9279	0.9292	0.9306	0.9319
1.5	0.9332	0.9345	0.9357	0.9370	0.9382	0.9394	0.9406	0.9418	0.9429	0.9441
1.6	0.9452	0.9463	0.9474	0.9484	0.9495	0.9505	0.9515	0.9525	0.9535	0.9545
1.7	0.9554	0.9564	0.9573	0.9582	0.9591	0.9599	0.9608	0.9616	0.9625	0.9633
1.8	0.9641	0.9649	0.9656	0.9664	0.9671	0.9678	0.9686	0.9693	0.9699	0.9706
1.9	0.9713	0.9719	0.9726	0.9732	0.9738	0.9744	0.9750	0.9756	0.9761	0.9767
2.0	0.9772	0.9778	0.9783	0.9788	0.9793	0.9798	0.9803	0.9808	0.9812	0.9817
2.1	0.9821	0.9826	0.9830	0.9834	0.9838	0.9842	0.9846	0.9850	0.9854	0.9857
2.2	0.9861	0.9864	0.9868	0.9871	0.9875	0.9878	0.9881	0.9884	0.9887	0.9890
2.3	0.9893	0.9896	0.9898	0.9901	0.9904	0.9906	0.9909	0.9911	0.9913	0.9916
2.4	0.9918	0.9920	0.9922	0.9925	0.9927	0.9929	0.9931	0.9932	0.9934	0.9936
2.5	0.9938	0.9940	0.9941	0.9943	0.9945	0.9946	0.9948	0.9949	0.9951	0.9952
2.6	0.9953	0.9955	0.9956	0.9957	0.9959	0.9960	0.9961	0.9962	0.9963	0.9964
2.7	0.9965	0.9966	0.9967	0.9968	0.9969	0.9970	0.9971	0.9972	0.9973	0.9974
2.8	0.9974	0.9975	0.9976	0.9977	0.9977	0.9978	0.9979	0.9979	0.9980	0.9981
2.9	0.9981	0.9982	0.9982	0.9983	0.9984	0.9984	0.9985	0.9985	0.9986	0.9986
3.0	0.9987	0.9987	0.9987	0.9988	0.9988	0.9989	0.9989	0.9989	0.9990	0.9990
3.1	0.9990	0.9991	0.9991	0.9991	0.9992	0.9992	0.9992	0.9992	0.9993	0.9993
3.2	0.9993	0.9993	0.9994	0.9994	0.9994	0.9994	0.9994	0.9995	0.9995	0.9995
3.3	0.9995	0.9995	0.9995	0.9996	0.9996	0.9996	0.9996	0.9996	0.9996	0.9997
3.4	0.9997	0.9997	0.9997	0.9997	0.9997	0.9997	0.9997	0.9997	0.9997	0.9998
3.5	0.9998	0.9998	0.9998	0.9998	0.9998	0.9998	0.9998	0.9998	0.9998	0.9998
3.6	0.9998	0.9998	0.9999	0.9999	0.9999	0.9999	0.9999	0.9999	0.9999	0.9999
3.7	0.9999	0.9999	0.9999	0.9999	0.9999	0.9999	0.9999	0.9999	0.9999	0.9999
3.8	0.9999	0.9999	0.9999	0.9999	0.9999	0.9999	0.9999	0.9999	0.9999	0.9999
3.9	1.0000	1.0000	1.0000	1.0000	1.0000	1.0000	1.0000	1.0000	1.0000	1.0000

Appendix III: Critical values – t-statistic and chi-squared, χ^2

Tails	Degrees of freedom	t-statistic Significance		Chi-squared Significance	
		95%	99%	95%	99%
1 or 2	df	0.05	0.01	0.05	0.01
2	1	12.71	63.66		
1		6.31	31.82	3.84	6.63
2	2	4.30	9.92		
1		2.92	6.96	5.99	9.21
2	3	3.18	5.84		
1		2.35	4.54	7.81	11.34
2	4	2.78	4.60		
1		2.13	3.75	9.49	13.28
2	5	2.57	4.03		
1		2.02	3.36	11.07	15.09
2	6	2.45	3.71		
1		1.94	3.14	12.59	16.81
2	7	2.36	3.50		
1		1.89	3.00	14.07	18.48
2	8	2.31	3.36		
1		1.86	2.90	15.51	20.09
2	9	2.26	3.25		
1		1.83	2.82	16.92	21.67
2	10	2.23	3.17		
1		1.81	2.76	18.31	23.21
2	11	2.20	3.11		
1		1.80	2.72	19.68	24.72
2	12	2.18	3.05		
1		1.78	2.68	21.03	26.22
2	13	2.16	3.01		
1		1.77	2.65	22.36	27.69
2	14	2.14	2.98		
1		1.76	2.62	23.68	29.14
2	15	2.13	2.95		
1		1.75	2.60	25.00	30.58
2	16	2.12	2.92		
1		1.75	2.58	26.30	32.00
2	17	2.11	2.90		
1		1.74	2.57	27.59	33.41
2	18	2.10	2.88		
1		1.73	2.55	28.87	34.81
2	19	2.09	2.86		
1		1.73	2.54	30.14	36.19
2	20	2.09	2.85		
1		1.72	2.53	31.41	37.57
2	25	2.06	2.79		
1		1.71	2.49	37.65	44.31
2	50	2.01	2.68		
1		1.68	2.40	67.50	76.15
2	Infinity	1.96	2.58		
1		1.64	2.33		

Appendix IV: Critical *F* values at 0.05 (95%) significance

Tails	dfN =															
		df_N = Degrees of freedom for Numerator														
		df_D = Degrees of freedom for Denominator														
	dfN =	1	2	3	4	5	6	7	8	9	10	15	20	30	50	100
1 or 2	dfD															
2	1	648	799	864	900	922	937	948	957	963	969	985	993	1001	1008	1013
1		161	199	216	225	230	234	237	239	241	242	246	248	250	252	253
2	2	38.5	39.0	39.2	39.2	39.3	39.3	39.4	39.4	39.4	39.4	39.4	39.4	39.5	39.5	39.5
1		18.5	19.0	19.2	19.2	19.3	19.3	19.4	19.4	19.4	19.4	19.4	19.4	19.5	19.5	19.5
2	3	17.4	16.0	15.4	15.1	14.9	14.7	14.6	14.5	14.5	14.4	14.3	14.2	14.1	14.0	14.0
1		10	10	9.28	9.12	9.01	8.94	8.89	8.85	8.81	8.79	8.70	8.66	8.62	8.58	8.55
2	4	12.2	10.6	9.98	9.60	9.36	9.20	9.07	8.98	8.90	8.84	8.66	8.56	8.46	8.38	8.32
1		7.71	6.94	6.59	6.39	6.26	6.16	6.09	6.04	6.00	5.96	5.86	5.80	5.75	5.70	5.66
2	5	10.0	8.43	7.76	7.39	7.15	6.98	6.85	6.76	6.68	6.62	6.43	6.33	6.23	6.14	6.08
1		6.61	5.79	5.41	5.19	5.05	4.95	4.88	4.82	4.77	4.74	4.62	4.56	4.50	4.44	4.41
2	6	8.81	7.26	6.60	6.23	5.99	5.82	5.70	5.60	5.52	5.46	5.27	5.17	5.07	4.98	4.92
1		5.99	5.14	4.76	4.53	4.39	4.28	4.21	4.15	4.10	4.06	3.94	3.87	3.81	3.75	3.71
2	7	8.07	6.54	5.89	5.52	5.29	5.12	4.99	4.90	4.82	4.76	4.57	4.47	4.36	4.28	4.21
1		5.59	4.74	4.35	4.12	3.97	3.87	3.79	3.73	3.68	3.64	3.51	3.44	3.38	3.32	3.27
2	8	7.57	6.06	5.42	5.05	4.82	4.65	4.53	4.43	4.36	4.30	4.10	4.00	3.89	3.81	3.74
1		5.32	4.46	4.07	3.84	3.69	3.58	3.50	3.44	3.39	3.35	3.22	3.15	3.08	3.02	2.97
2	9	7.21	5.71	5.08	4.72	4.48	4.32	4.20	4.10	4.03	3.96	3.77	3.67	3.56	3.47	3.40
1		5.12	4.26	3.86	3.63	3.48	3.37	3.29	3.23	3.18	3.14	3.01	2.94	2.86	2.80	2.76
2	10	6.94	5.46	4.83	4.47	4.24	4.07	3.95	3.85	3.78	3.72	3.52	3.42	3.31	3.22	3.15
1		4.96	4.10	3.71	3.48	3.33	3.22	3.14	3.07	3.02	2.98	2.85	2.77	2.70	2.64	2.59
2	15	6.20	4.77	4.15	3.80	3.58	3.41	3.29	3.20	3.12	3.06	2.86	2.76	2.64	2.55	2.47
1		4.54	3.68	3.29	3.06	2.90	2.79	2.71	2.64	2.59	2.54	2.40	2.33	2.25	2.18	2.12
2	20	5.87	4.46	3.86	3.51	3.29	3.13	3.01	2.91	2.84	2.77	2.57	2.46	2.35	2.25	2.17
1		4.35	3.49	3.10	2.87	2.71	2.60	2.51	2.45	2.39	2.35	2.20	2.12	2.04	1.97	1.91
2	30	5.57	4.18	3.59	3.25	3.03	2.87	2.75	2.65	2.57	2.51	2.31	2.20	2.07	1.97	1.88
1		4.17	3.32	2.92	2.69	2.53	2.42	2.33	2.27	2.21	2.16	2.01	1.93	1.84	1.76	1.70
2	50	5.34	3.97	3.39	3.05	2.83	2.67	2.55	2.46	2.38	2.32	2.11	1.99	1.87	1.75	1.66
1		4.03	3.18	2.79	2.56	2.40	2.29	2.20	2.13	2.07	2.03	1.87	1.78	1.69	1.60	1.52
2	100	5.18	3.83	3.25	2.92	2.70	2.54	2.42	2.32	2.24	2.18	1.97	1.85	1.71	1.59	1.48
1		3.94	3.09	2.70	2.46	2.31	2.19	2.10	2.03	1.97	1.93	1.77	1.68	1.57	1.48	1.39

Example

Critical *F*-value for one-tailed test with $df_N = 15$ and $df_D = 8$ at 0.05 significance:

$$F_{1,0.05,15,8} = 3.22$$

Appendix V: Critical values at 0.05 (95%) significance for Pearson's correlation coefficient, r, Spearman's Rank correlation coefficient, r_S, and Wilcoxon lower limit, W_L

Tails	Sample size	Pearson's	Spearman's	Wilcoxon
1 or 2	n	r	r_S	W_L
2	4	0.950		
1		0.900		
2	5	0.878	-	
1		0.805	0.900	
2	6	0.811	0.886	-
1		0.729	0.829	2
2	7	0.754	0.786	2
1		0.669	0.714	3
2	8	0.707	0.738	3
1		0.622	0.643	5
2	9	0.666	0.683	5
1		0.582	0.600	8
2	10	0.632	0.648	8
1		0.549	0.564	10
2	11	0.602	0.623	10
1		0.521	0.523	13
2	12	0.576	0.591	13
1		0.497	0.497	17
2	13	0.553	0.566	17
1		0.476	0.475	21
2	14	0.532	0.545	21
1		0.456	0.457	25
2	15	0.514	0.525	25
1		0.441	0.441	30
2	16	0.497	0.507	29
1		0.426	0.425	35
2	17	0.482	0.490	34
1		0.412	0.412	41
2	18	0.468	0.476	40
1		0.400	0.399	47
2	19	0.456	0.462	46
1		0.389	0.388	53
2	20	0.444	0.450	52
1		0.378	0.377	60
2	25	0.396	0.400	89
1		0.337	0.336	100
2	30	0.361	0.364	137
1		0.306	0.305	151

Appendix VI: Mann–Whitney lower limit, U_L, at 0.05 (95%) significance

Tails (1 or 2)	n_2	$n_1=2$	3	4	5	6	7	8	9	10	11	12	13	14	15	16	17	18	19	20
2	2										0	1	1	1	1	1	2	2	2	2
1							0	1	1	1	1	2	2	3	3	3	3	4	4	4
2	3				0	1	1	2	2	3	3	4	4	5	5	6	6	7	7	8
1				0	1	2	2	3	4	4	5	5	6	7	7	8	9	9	10	11
2	4			0	1	2	3	4	4	5	6	7	8	9	10	11	11	12	13	14
1			0	1	2	3	4	5	6	7	8	9	10	11	12	14	15	16	17	18
2	5		0	1	2	3	5	6	7	8	9	11	12	13	14	15	17	18	19	20
1			1	2	4	5	6	8	9	11	12	13	15	16	18	19	20	22	23	25
2	6		1	2	3	5	6	8	10	11	13	14	16	17	19	21	22	24	25	27
1			2	3	5	7	8	10	12	14	16	17	19	21	23	25	26	28	30	32
2	7		1	3	5	6	8	10	12	14	16	18	20	22	24	26	28	30	32	34
1		0	2	4	6	8	11	13	15	17	19	21	24	26	28	30	33	35	37	39
2	8	0	2	4	6	8	10	13	15	17	19	22	24	26	29	31	34	36	38	41
1		1	3	5	8	10	13	15	18	20	23	26	28	31	33	36	39	41	44	47
2	9	0	2	4	7	10	12	15	17	20	23	26	28	31	34	37	39	42	45	48
1		1	4	6	9	12	15	18	21	24	27	30	33	36	39	42	45	48	51	54
2	10	0	3	5	8	11	14	17	20	23	26	29	33	36	39	42	45	48	52	55
1		1	4	7	11	14	17	20	24	27	31	34	37	41	44	48	51	55	58	62
2	11	0	3	6	9	13	16	19	23	26	30	33	37	40	44	47	51	55	58	62
1		1	5	8	12	16	19	23	27	31	34	38	42	46	50	54	57	61	65	69
2	12	1	4	7	11	14	18	22	26	29	33	37	41	45	49	53	57	61	65	69
1		2	5	9	13	17	21	26	30	34	38	42	47	51	55	60	64	68	72	77
2	13	1	4	8	12	16	20	24	28	33	37	41	45	50	54	59	63	67	72	76
1		2	6	10	15	19	24	28	33	37	42	47	51	56	61	65	70	75	80	84
2	14	1	5	9	13	17	22	26	31	36	40	45	50	55	59	64	69	74	78	83
1		3	7	11	16	21	26	31	36	41	46	51	56	61	66	71	77	82	87	92
2	15	1	5	10	14	19	24	29	34	39	44	49	54	59	64	70	75	80	85	90
1		3	7	12	18	23	28	33	39	44	50	55	61	66	72	77	83	88	94	100
2	16	1	6	11	15	21	26	31	37	42	47	53	59	64	70	75	81	86	92	98
1		3	8	14	19	25	30	36	42	48	54	60	65	71	77	83	89	95	101	107
2	17	2	6	11	17	22	28	34	39	45	51	57	63	69	75	81	87	93	99	105
1		3	9	15	20	26	33	39	45	51	57	64	70	77	83	89	96	102	109	115
2	18	2	7	12	18	24	30	36	42	48	55	61	67	74	80	86	93	99	106	112
1		4	9	16	22	28	35	41	48	55	61	68	75	82	88	95	102	109	116	123
2	19	2	7	13	19	25	32	38	45	52	58	65	72	78	85	92	99	106	113	119
1		4	10	17	23	30	37	44	51	58	65	72	80	87	94	101	109	116	123	130
2	20	2	8	14	20	27	34	41	48	55	62	69	76	83	90	98	105	112	119	127
1		4	11	18	25	32	39	47	54	62	69	77	84	92	100	107	115	123	130	138

Short Answers to 'Q' Questions

The answers below give the numerical answer to most of the 'Q' questions. Full worked answers on *video* are available on the Website. Some answers, given as 'Web' below, are only available on the Website.

Chapter 2

Q2.1 4.26×10^4, 3.62×10^{-3}, 1.0×10^4, 1.0×10^{-4}, 4.5×10^2, 2.66×10^4, 3.2×10^3, 4.5×10^{-6}

Q2.2 3.6×10^2, 1.5×10^3, 4.0×10^1, 5.0×10^4, 2.24×10^1, 4.0×10^7

Q2.3 1.2042×10^3, 2.3771×10^9

Q2.4 0.047, 0.046, 14.0, 7.35×10^3, 27000, 11.3, 11.2, 5.65×10^{-3}

Q2.5 0047, 8.00, 426.89, 1.35

Q2.6 3.89 g

Q2.7 500 mph

Q2.8 4.0×10^{-5}, 2083.33, 0.1555, 12.159, ± 0.259

Q2.9 12

Q2.10 3150 kJ, 285.7 kcal, 65.9 kg, 152.4 mm by 25.4 mm, 22.72 L, 11.5 g

Q2.11 100, 7.9×10^3 kg m^{-3}, 0.15 kg m^{-2}, 7.06 L per 100 km

Q2.12 10.80

Q2.13 180, 180 g mol^{-1}, 180 g

Q2.14 106 g, 15.9 g, 0.033 mol

Q2.15 122.2 g mol^{-1}, 122.2

Q2.16 0.01 L, 1.16×10^{-5} L, 67 mL, 0.26 μL

Q2.17 0.2 mol L^{-1}

Q2.18 1.867 mol L^{-1}

Q2.19 0.292 g

Q2.20 6.2425 g

Q2.21 0.1 mol L^{-1}

Q2.22 5 mL

Q2.23 20 mL

Q2.24 2π (= 6.283), 0.5π (= 1.57), 2.967, 57.30°, 120.32°, 630°

Q2.25 6.98 cm

Q2.26 8.66 m

Q2.27 149.67 m

Q2.28 Web

Q2.29 3840 km
Q2.30 36.87°
Q2.31 D1*TAN(RADIANS(D3))

Chapter 3

Q3.1   12, 36, 4, 0.222, −4, 36, 0, 2.5
Q3.2   17 m s^{-1}
Q3.3   0 m s^{-1}
Q3.4   F, T, T, T
Q3.5 $6 + 3x, 6 + 12x, 6{-}8x, px + 2p, -6p + 3px, px + p^2$
Q3.6 $v^2 - 4v + 4, v^2 - t^2, 3v - tv^2 - 4vt + 12, 3xv + vy + 3tx + ty, px - qx - 2py + 2qy, 5v + v^2 - t^2 + 6 - t$
Q3.7   9, 7, 10, [2,6,11]
Q3.8 $A = (1/2)bh, d = vt, BMI = m/(h^2), V = (4/3)\pi r^3, v = \sqrt{2gh}$
Q3.9   4 and $(x + y)$, 2 and $(2x + 3y)$, 2 and x and $(2 + 3x)$, p and x and $(q + b)$
Q3.10 $2a/b, 2a/b, (4a + 1)/2b, (2a + 3)/2b$
Q3.11 Y, Y, N, Y, Y, Y
Q3.12 $8k + 2x = 16, 2x = 16{-}8k, x = 8 - 4k$, Yes
Q3.13 $(x/4) = 3 - 2y, x = 4(3 - 2y) = 12{-}8y$, Yes
Q3.14 $x = 2p - 4, x = -3(2 - \mu) + 4p, x = -8 + (q - t)$
Q3.15 $x = p - 18, x = 10{-}3(q - p), x = 9 + p, x = (2v - t) - 8 + k$
Q3.16 $x = 30 + 10p, x = 20{-}8q, x = k/(s - m), x = 16p - 8q$
Q3.17 1, −2, 4, 3, 5, 14
Q3.18 $x = p - 6, x = 7 - k, x = 3, x = p - 8$
Q3.19 Web
Q3.20 R is fundamental constant, p and n are constants for Charles' law, V and T are variables
Q3.21 Web
Q3.22 ±5, ±0.158, 81, 19.95, 3.27, 19.88, 2.76, −2
Q3.23 $x = 20, x = -2, x = (5 - k)/p, x = (2q - v)/q$
Q3.24 $x = \sqrt{(25 - t)}, x = \sqrt{25} - t, x = [\sin^{-1}(0.6k)]/2, x = 4/(a - y), x = \ln(y)/2, x = y^2/2, x = y^2/18 + 7/2, x = 2/(y^2 - k)$
Q3.25 198.1, 79.24, 3.258, −0.0464, 0.809, 0.39
Q3.26 Web
Q3.27 $x = -6, x = (3 + a)/3, x = (8 + 3p)/4, x = 2p/3, x = 3a/2, x = 11/5$
Q3.28 $3, 5p, (m - 3), (3 - p), -2 - p, p + 2k - 1$
Q3.29 $x + 10, 9x + 8, 10x + 2, 2x(b - a), (a + 5)x - 3a, x$
Q3.30 $x = 1.5, x = -(a + b)/2, x = 1.5, x = 5/(3 - p), x = (5 - p)/(3 - p), x = (5 - p)/(p - q - 2), x = 12a/(6a - 1), x = (2p - 6)/6, x = (2p - 6)/(p - 1)$
Q3.31 Web
Q3.32 1.0 or 0.5, 1.025 or −0.244, −2.87 or 0.871, 5.24 or 0.764
Q3.33 Equal roots $= 2$, 1 or −1.5, no real solutions, equal roots $= -1.5$
Q3.34 1285.7 s, 2042.9 m
Q3.35 $C_1 = 0.210, C_2 = 0.150$
Q3.36 (1,1) or (−1.5,2.25)
Q3.37 Web

Chapter 4

Q4.1 T, F, F

Q4.2 $[-1, 1, 3, 5, 7]$, Web, 3, N, Y, N, Web

Q4.3 $L = 2.299 \times 10^{-5} \times T + 1.21$, 2.299×10^{-5}, 1.21

Q4.4 $-0.75, -0.5$

Q4.5 7/4

Q4.6 $T = 40W + 20$, 2.5 kg

Q4.7 (c), 7 s, -260 m

Q4.8 2, 2/3, -4, -0.5, -1, 6.5

Q4.9 $y = 2x + 3$, $y = 0.5x + 2$, $y = -0.5x + 4$

Q4.10 $F = 1.8C + 32$, -17.78

Q4.11 $y = 3x - 2$

Q4.12 $y = -(1/3)x + (4/3)$

Q4.13 $x = -1.5$

Q4.14 3.9, 7.2, 2, -4

Q4.15 16.25, 22

Q4.16 0.125, 7.125°

Q4.17 $y =$ Heart rate, $x =$ Step$-$up rate

Q4.18 C, R, C, C

Q4.19 $P = 1.235Q + 2.845$, 7.044

Q4.20 0.0308, 0.051

Q4.21 0.03344

Q4.22 $\theta = 0.8788w$, 5.58 g

Q4.23 No, Yes, RT, 0, Yes, 8.33 J K^{-1} mol^{-1}, Slope increases

Q4.24 $y = 1/v$, $x = 1/S$, $v_{\max} = 1/c$, $K_M = m/c$

Chapter 5

Q5.1 $10^5, 10^6, 10, 10^5, e^2, e^5, e^2, e^6$

Q5.2 2511.9, 0.3548, 1, 0.657, 2.718, 1

Q5.3 2.54, 1.54, 0.54, 2, -2, -0.301, 0.301, 1.301, 1, 2.303

Q5.4 3.09, 0.828, 2.888, 5.366, -1.788, 1.255, 0.628, 7, 5.01×10^{-10}, -609.57

Q5.5 -0.3, 0.62, 0.3, 1.3, -0.3, 6.9, 0.69, 0.93

Q5.6 0.799, 0.588, 2.52, 0.748

Q5.7 (i) Plot $\log(V)$ against $\log(A)$, slope $m = k$ and intercept $c = \log(p)$, so $p = 10^c$
(ii) Plot $\log(E)$ against $\log(T + 273)$, slope $m = z$ and intercept $c = \log(\sigma)$, so $\sigma = 10^c$

Q5.8 5.0×10^6, 5.0×10^4

Q5.9 73 dB, 79 dB, 90 dB, 67 dB, 61 dB, 50 dB

Q5.10 8.47, 3.47

Q5.11 6.31×10^{-10} mol L^{-1}, 6.31×10^{-4} mol L^{-1}

Q5.12 0.1 %, 10 %, 100 %, 2, 0.3

Q5.13 Web

Q5.14 2, 1.5, 0.9, 0.1

Q5.15 2 weeks, 6, 19 weeks

Q5.16 2.0×10^3, 0.1, 2.936×10^4

Q5.17 $1.35 \text{ s}^{-1}, 1.743 \text{ s}^{-1}$
Q5.18 28 minutes
Q5.19 7.825 hours
Q5.20 0.0152 s (or 15.2 ms)
Q5.21 $N_t = 450 \exp(0.0320 \times t)$, 661
Q5.22 5771, 9514, 15686
Q5.23 $N_t = 100 \exp(0.0953t)$, $N_0 = 100$, $k = 0.0953$
Q5.24 Web, 0.632, 0.865
Q5.25 1.257 lumens
Q5.26 57.75 s
Q5.27 $K = 0.61 \text{ h}^{-1}, C_0 = 60.3$

Chapter 6

Q5.1 5000, 300, -2000, -10
Q6.2 0, 0.25, 0.75, 1.5 , 2, 1.75, 1, 0.25, 0, -0.25, -0.5, -0.5, -0.25, 0
Q6.3 Estimated values: 9350 cells per minute, 19 600 cells per minute
Q6.4 $N_t = N_0 e^{kt}$
Q6.5 $2.24 \times 10^8, 3.03 \times 10^7, 4.10 \times 10^6$
Q6.6 6.22×10^{-3} (Web)

Chapter 7

Q7.1 (i) 6.6, 5.35, 7.75, 2.4 (ii) 17, 10.5, 25.25, 14.75 (iii) 56, 34, 72.5, 38.5
Q7.2 Web
Q7.3 Web
Q7.4 P, S, S, P, P, S
Q7.5 24.0
Q7.6 (i) 4.8 (ii) 0.2325
Q7.7 (i) 4.8 (ii) 0.1, -0.1, 0.3, 0.1, -0.4 (iii) 0 (iv) Yes (v) 0.28 (vi) 0.07 (vii) 0.0265
Q7.8 (i) 6.4, 2.3, 1.517 (ii) 66.4, 2.3, 1.517 (iii) 416.4, 2.3, 1.517
Q7.9 (i) Web (ii) 2, 3, 10, 5, 4
Q7.10 (i) 4, 10, 7, 3 (ii) 0.167, 0.417, 0.292, 0.125 (iii) 1.001 (iv) Web
Q7.11 (i) Web (ii) 0.533, 0.057, 0.324, 0.086 (iii) 192, 21, 117, 31 (iv) Web
Q7.12 (i) 10, 4, 2, 2, 4,10 (ii) 0.6, 5.25, 8.5, 7.5, 4, 0.9 (iii) Web
Q7.13 (i) 4, 10, 7, 3 (ii) 14, 18, 22, 26 (iii) 56, 180, 154, 78 (iv) 468 (v) 19.5 (vi) £461 500
        (vii) £19 229
Q7.14 (i) 1, 2, 3, 4, 5, 6, 5, 4, 3, 2, 1 (ii) Web (iii) 36
Q7.15 (i) 0.0278, 0.0556, 0.0833, etc. (ii) Web (iii) 1.00
Q7.16 0.05
Q7.17 (i) 0.0278, 0.0833, 0.1667, etc. (ii) 0.4167 (iii) 0.1667
Q7.18 (i) 0.243, 0.357, 0.271, 0.129 (ii) 0.129 (iii) Web (iv) 7, 11, 8, 4
Q7.19 1/52, 1/13, 3/13, 1/13, 1/4
Q7.20 9/16, 3/16, 3/16, 1/16
Q7.21 7/24, 15/24, 1
Q7.22 1/26, 1/13, 4/13

Q7.23   1/13, 12/13, 1
Q7.24   1/36, 1/18, 1/6, 1/36
Q7.25   0.9, 0.1, 0.6561, 0.0001, 0.0036, 0.2916
Q7.26   0.0065
Q7.27   1/7, 2/7, 2/7, 2/7, 1
Q7.28   0.618
Q7.29   120, 1, 1, 0, 42, 10100
Q7.30   24
Q7.31   1680
Q7.32   167 960

Chapter 8

Q8.1   0.1915, 0.2417, 0.5000, 0.6828, 0.9546, 0.9974, 1.000
Q8.2   1.5, 1.2, 1.2
Q8.3   68 300, 95 000, 300, 15 850
Q8.4   Web
Q8.5   19.08 to 26.92
Q8.6   1.0, 21.04 to 24.96
Q8.7   16.8 to 21.9, 17.9 to 20.0, Web
Q8.8   0.08 ppm
Q8.9   1.87, 0.21
Q8.10   10.75, 0.5
Q8.11   0.2373, 0.3955, 0.2637, 0.0879, 0.0146, 0.0010, 1.0000
Q8.12 5.9×10^{-6}, 1.4×10^{-4}, 0.233, 0.121, 0.028, 0.383, 0.617
Q8.13   0.606, 0.303, 0.076, 0.014
Q8.14 0.74 ± 0.12
Q8.15   (i) $CD_A = 0.99$, $CD_B = 1.33$ (ii) A (iii) Clumping

Chapter 9

Q9.1   1, 2, 1, 2
Q9.2   F, T, F, F

Chapter 10

Q10.1   (ii) 0.05 (iii) 3.585×10^{-8}, 0.080×10^{-8} (iv) 2.60 (v) 1 (vi) 5 (vii) 2.02
Q10.2   Web
Q10.3   (i) No (ii) c (iii) c
Q10.4   Web
Q10.5   Paired t-test
Q10.6 $t_{STAT} = 2.54$, $t_{CRIT} = 2.45$, Sig., $p = 0.044$ (Web)
Q10.7   3.68, 3.77
Q10.8   Web
Q10.9   Web

Chapter 11

Q11.1   0.094
Q11.2   Web
Q11.3   Web
Q11.4   Web
Q13.5   Web
Q13.6   Web

Chapter 12

Q12.1 $W+ = 92$, $W- = 28$, $W_{\mathrm{L}} = 30$, Sig.
Q12.2 $U_{\mathrm{Y}} = 12$, $U_{\mathrm{L}} = 12$, Sig., $p = 0.044$
Q12.3 $W(-) = 3$, $W_{\mathrm{L}} = 2$, Not sig.
Q12.4 $H = 7.43$, $\chi^2_{\mathrm{CRIT}} = 7.81$, Not sig.
Q12.5 $S = 9.53$, $\chi^2_{\mathrm{CRIT}} = 7.81$, Sig.
Q12.6 $S = 7.73$, $\chi^2_{\mathrm{CRIT}} = 7.81$, Not sig. (Web)

Chapter 13

Q13.1   (i) $r = 0.715$ (ii) $r_{\mathrm{crit}} = 0.622$, Sig. (iii) $p = 0.023$
Q13.2   (i) $r = 0.542$, (ii) $r_{\mathrm{crit}} = 0.549$, Not sig. (iii) $p = 0.0526$
Q13.3   Web
Q13.4   1.578, 0.1477, 1.11 to 2.05

Chapter 14

Q14.1   9.49, 6.63
Q14.2 $\chi^2_{\mathrm{CRIT}} = 7.81$, $\chi^2_{\mathrm{STAT}} = 3.6$, No
Q14.3 $\chi^2_{\mathrm{STAT}} = 5.23$, $\chi^2_{\mathrm{CRIT}} = 5.99$, No
Q14.4 $A : \chi^2_{\mathrm{STAT}} = 10.61$, $B : \chi^2_{\mathrm{STAT}} = 2.06$, $\chi^2_{\mathrm{CRIT}} = 9.49$, Web
Q14.5   (ii) $\chi^2_{\mathrm{STAT}} = 3.42$ (iii) $\chi^2_{\mathrm{STAT}} = 4.05$, $\chi^2_{\mathrm{CRIT}} = 3.84$
Q14.6 $\chi^2_{\mathrm{STAT}} = 9.5$, $\chi^2_{\mathrm{CRIT}} = 11.07$, No
Q14.7 $\chi^2_{\mathrm{STAT}} = 12$, $\chi^2_{\mathrm{CRIT}} = 7.81$, Yes
Q14.8 $\chi^2_{\mathrm{STAT}} = 3.28$, $\chi^2_{\mathrm{CRIT}} = 5.99$, No
Q14.9   0.0284, 0.0568, symmetrical distribution
Q14.10   (i) 0.504 to 0.696 (ii) biased
Q14.11   Yes, if the Yates correction used

Index

Note: Figures and Tables are indicated by *italic* page numbers